METHODS IN MOLECULAR BIOLOGY™

Series Editor
John M. Walker
School of Life Sciences
University of Hertfordshire
Hatfield, Hertfordshire, AL10 9AB, UK

For further volumes:
http://www.springer.com/series/7651

Chemical Neurobiology

Methods and Protocols

Edited by

Matthew R. Banghart

Department of Neurobiology, Harvard Medical School, Boston, MA, USA

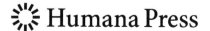 Humana Press

Editor
Matthew R. Banghart
Department of Neurobiology
Harvard Medical School
Boston, MA, USA

ISSN 1064-3745   ISSN 1940-6029 (electronic)
ISBN 978-1-62703-344-2   ISBN 978-1-62703-345-9 (eBook)
DOI 10.1007/978-1-62703-345-9
Springer New York Heidelberg Dordrecht London

Library of Congress Control Number: 2013932702

Printed on acid-free paper

Humana Press is a brand of Springer
Springer is part of Springer Science+Business Media (www.springer.com)

Preface

Many advances in modern neuroscience are enabled by the availability of chemical tools that allow sensitive, precise, and quantitative measurements of, and control over, biological processes. These powerful reagents are widely used for investigating the nervous system at levels of detail ranging from ion channel structure to neural network dynamics. Recent advances in photochemistry, microscopy, and protein engineering have triggered a surge in the development and application of these interdisciplinary techniques. Chemical Neurobiology: Methods and Protocols is intended to assist with the design, characterization, and validation of new chemical tools for neurobiology by providing detailed protocols of procedures and assays deemed essential for their successful development and implementation.

The methods covered are divided into three parts: chemical probes of membrane protein structure and function, photochemical control of protein and cellular function, and chemical probes for imaging in the nervous system. The first part addresses several emerging approaches to neurochemical pharmacology, which not only shed light on ligand–receptor interactions but also provide insight into the delicate relationships between protein structure and function. These protocols cover the use of unnatural amino acids and covalent peptide-toxins to study ion channel structure and function, a novel high-throughput genetic screen that yields insight into small molecule–ion channel interactions, and the application of a functional calcium imaging assay to probe the unique pharmacological properties of bivalent GPCR ligands with heteromeric receptors. As modern neuroscience relies heavily on optical methodology, the remainder of the book is devoted to protocols that will guide the characterization of photochemical reagents that enable researchers to control and detect molecular and cellular signaling. In the context of photopharmacology, the protocols presented will guide the quantification of key properties of caged neurotransmitters such as quantum yield and photolysis kinetics, the wavelength sensitivity and mechanistic pharmacology of photoswitchable ligands for ion channels, and principles underlying the design and implementation of photoreactive ligands for neurotransmitter receptors. On the topic of sensors, this book includes assays for determining the affinity and fluorescence properties of small molecule ion sensors, the sensitivity and optimal optical parameters for imaging membrane potential with voltage-sensitive dyes, pharmacological characterization and optical implementation of fluorescent ligands for neurotransmitter receptors, the use of quantum dots for measuring vesicular release of neurotransmitters, and the innovative application of directed evolution to create protein-based probes for neurotransmitters that can be monitored in vivo by MRI.

Instructing by example, each protocol includes background on the biological problems addressable by the tool or technique and considerations critical to the initial design process. Importantly, each protocol is written in a format that can be readily extended to related systems in order to facilitate the development of new chemical tools.

The topics covered should be of value to scientists at many levels, including students aiming to expand their perspective, laboratory researchers seeking technical guidance, and established investigators looking for creative solutions to their research problems in molecular, cellular, and systems neuroscience.

Boston, MA, USA   *Matthew R. Banghart*

About This Book

Chemical Neurobiology provides powerful chemical tools and techniques that are widely used for investigating the nervous system at levels of detail ranging from ion channel structure to neural network analysis. Recent advances in microscopy, protein engineering, and proteomics have triggered a surge in the development of these important techniques. This volume is intended to assist in the design, characterization, and validation of new chemical tools for neurobiology by providing protocols of assays deemed essential for the successful development of existing and emerging techniques. Instructing by example, each protocol includes background on the biological problems addressable by the technique and considerations critical to the initial design process. Importantly, each protocol is written in a format that can be readily extended to related systems in order to facilitate the development of new chemical tools. Chapters consist of detailed step-by-step procedures that include reagents, equipment, technical diagrams, and extensive discussion of practical considerations as well as common pitfalls and their solutions. This book targets scientists at many levels, including undergraduates seeking to expand their perspective, graduate students, postdocs and researchers requiring experimental guidance, and established investigators looking to develop new solutions to their research problems.

Written for:
Chemists, chemical biologists, biochemists, pharmacologists, molecular and cellular biologists, physiologists, and neuroscientists.

Contents

PART III   CHEMICAL PROBES FOR IMAGING IN THE NERVOUS SYSTEM

Contributors

COREY D. ACKER • *Center for Cell Analysis and Modeling, University of Connecticut Health Center, Farmington, CT, USA*

FRANCES H. ARNOLD • *Division of Chemistry and Chemical Engineering, California Institute of Technology, Pasadena, CA, USA*

SVIATOSLAV N. BAGRIANTSEV • *Cardiovascular Research Institute, University of California San Francisco, San Francisco, CA, USA*

MATTHEW R. BANGHART • *Department of Neurobiology, Harvard Medical School, Boston, MA, USA*

FREDERIC BOLZE • *Department of Bioorganic Chemistry, Centre National de la Recherche Scientifique, Université Louis Pasteur Strasbourg, Illkirch, France*

LAUREL A. COOLEY • *Molecular and Cell Biology Division, Life Technologies, Eugene, OR, USA*

COREY A. COSTANTINO • *Department of Physiology and Pharmacology, Oregon Health and Science University, Portland, OR, USA*

PAMELA M. ENGLAND • *Department of Pharmaceutical Chemistry, University of California San Francisco, San Francisco, CA, USA; Department of Cellular and Molecular Pharmacology, University of California San Francisco, San Francisco, CA, USA*

TIMM FEHRENTZ • *Department of Chemistry, Ludwig-Maximillians-Universität-University of Munich, Munich, Germany; Department of Pharmacology, Ludwig-Maximillians-Universität-University of Munich, Munich, Germany; Center for Integrated Protein Science, Ludwig-Maximillians-Universität-University of Munich, Munich, Germany; Department of Molecular and Cellular Biology, University of California Berkeley, Berkeley, CA, USA*

KYLE R. GEE • *Molecular and Cell Biology Division, Life Technologies, Eugene, OR, USA*

MAURICE GOELDNER • *Department of Bioorganic Chemistry, Centre National de la Recherche Scientifique, Université Louis Pasteur Strasbourg, Illkirch, France*

JESSICA H. HARVEY • *Ernest Gallo Clinic and Research Center, University of California San Francisco, Emeryville, CA, USA*

RUUD HOVIUS • *Laboratory of Physical Chemistry of Polymers and Membranes, Institute of Chemical Sciences and Engineering, Swiss Federal Institute of Technology (EPFL), Lausanne, Switzerland*

ZHENGMAO HUA • *Department of Biochemistry and Molecular Pharmacology, University of Massachusetts Medical School, Worcester, MA, USA*

ALAN JASANOFF • *Department of Biological Engineering, Massachusetts Institute of Technology, Cambridge, MA, USA; Department of Brain and Cognitive Sciences, Massachusetts Institute of Technology, Cambridge, MA, USA; Department of Nuclear Science and Engineering, Massachusetts Institute of Technology, Cambridge, MA, USA*

JOSEPH P.Y. KAO • *Center for Biomedical Engineering and Technology, University of Maryland School of Medicine, Baltimore, MD, USA; Department of Physiology, University of Maryland School of Medicine, Baltimore, MD, USA*

WILLIAM R. KOBERTZ • *Department of Biochemistry and Molecular Pharmacology, University of Massachusetts Medical School, Worcester, MA, USA*

ALEXANDER G. KOMAROV • *Department of Physiology and Pharmacology, Oregon Health and Science University, Portland, OR, USA*

RICHARD H. KRAMER • *Department of Molecular and Cellular Biology, University of California Berkeley, Berkeley, CA, USA*

LESLIE M. LOEW • *Center for Cell Analysis and Modeling, University of Connecticut Health Center, Farmington, CT, USA*

VLADIMIR V. MARTIN • *Molecular and Cell Biology Division, Life Technologies, Eugene, OR, USA*

DANIEL L. MINOR JR. • *Cardiovascular Research Institute, University of California San Francisco, San Francisco, CA, USA; Department of Biochemistry and Biophysics, University of California San Francisco, San Francisco, CA, USA; Department of Cellular and Molecular Pharmacology, University of California San Francisco, San Francisco, CA, USA; Physical Biosciences Division, Lawrence Berkeley National Laboratory, Berkeley, CA, USA*

ALEXANDRE MOUROT • *Department of Molecular and Cellular Biology, University of California Berkeley, Berkeley, CA, USA*

SUKUMARAN MURALIDHARAN • *Center for Biomedical Engineering and Technology, University of Maryland School of Medicine, Baltimore, MD, USA; Department of Physiology, University of Maryland School of Medicine, Baltimore, MD, USA*

JEAN FRANCOIS NICOUD • *Department of Bioorganic Chemistry, Centre National de la Recherche Scientifique, Université Louis Pasteur Strasbourg, Illkirch, France*

PHILIP A. ROMERO • *Division of Chemistry and Chemical Engineering, California Institute of Technology, Pasadena, CA, USA*

MIKHAIL G. SHAPIRO • *Department of Biological Engineering, Massachusetts Institute of Technology, Cambridge, MA, USA*

ALEXANDRE SPECHT • *Department of Bioorganic Chemistry, Centre National de la Recherche Scientifique, Université Louis Pasteur Strasbourg, Illkirch, France*

DIRK TRAUNER • *Department of Chemistry, Ludwig-Maximillians-Universität-University of Munich, Munich, Germany; Center for Integrated Protein Science, Ludwig-Maximillians-Universität-University of Munich, Munich, Germany*

FRANCIS I. VALIYAVEETIL • *Department of Physiology and Pharmacology, Oregon Health and Science University, Portland, OR, USA*

RICHARD M. VAN RIJN • *Ernest Gallo Clinic and Research Center, University of California San Francisco, Emeryville, CA, USA*

JENNIFER L. WHISTLER • *Ernest Gallo Clinic and Research Center, University of California San Francisco, Emeryville, CA, USA*

QI ZHANG • *Department of Pharmacology, Vanderbilt University, Nashville, TN, USA*

Part I

Chemical Probes of Membrane Protein Structure and Function

Chapter 1

Engineering K⁺ Channels Using Semisynthesis

Alexander G. Komarov, Corey A. Costantino, and Francis I. Valiyaveetil

Abstract

Potassium channels conduct K⁺ ions selectively and at very high rates. Central to the function of K⁺ channels is a structural unit called the selectivity filter. In the selectivity filter, a row of four K⁺ binding sites are created using mainly the backbone carbonyl oxygen atoms. Due to the involvement of the protein backbone, site-directed mutagenesis is of limited utility in investigating the selectivity filter. In order to overcome this limitation, we have developed a semisynthetic approach, which permits the use of chemical synthesis to manipulate the selectivity filter. In this chapter, we describe the protocols that we have developed for the semisynthesis of the K⁺ channel, KcsA. We anticipate that the protocols described in this chapter will also be applicable for the semisynthesis of other integral membrane proteins of interest.

Key words K⁺ channels, Membrane proteins, Semisynthesis, Solid-phase peptide synthesis

---

1 Introduction

Protein structure–function studies require the ability to modify the protein. This has been mainly carried out using traditional site-directed mutagenesis (SDM). SDM allows substitution of the residue of interest with the other naturally occurring amino acids. The nature of the modifications that can be introduced by SDM is therefore limited by the set of naturally occurring amino acids. Another limitation of SDM is that it cannot be used to manipulate the protein backbone. An alternate approach to protein modification is through the use of chemical synthesis. The advantage of chemical synthesis is that it enables the incorporation of a large number of unnatural amino acids, which allows the precise modification of the structural and electronic properties of the amino acid side chain. In addition, chemical synthesis enables us to manipulate the protein backbone.

We have relied on chemical synthesis in our investigations of K⁺ channels. K⁺ channels are integral membrane proteins that provide pathways for K⁺ ions to pass across biological membranes

Matthew R. Banghart (ed.), *Chemical Neurobiology: Methods and Protocols*, Methods in Molecular Biology, vol. 995,
DOI 10.1007/978-1-62703-345-9_1, © Springer Science+Business Media New York 2013

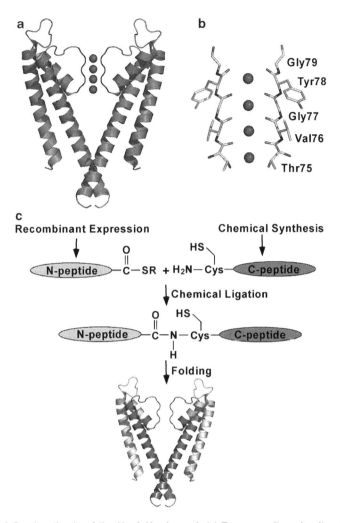

Fig. 1 Semisynthesis of the KcsA K+ channel. (**a**) Two opposite subunits of the tetrameric KcsA channel are shown (pdb:1k4c). The selectivity filter, residues 75–79, is colored *dark gray*. K+ ions bound to the selectivity filter are depicted as *spheres*. (**b**) Close-up view of the selectivity filter in stick representation with the front and the back subunits removed. (**c**) Semisynthetic strategy. The KcsA polypeptide (residues 1–123) is obtained by the expressed protein ligation of a recombinantly expressed N-peptide thioester (1–69, *light gray*) and a chemically synthesized peptide (70–123, *dark gray*) with an N-terminal cysteine. The KcsA polypeptide obtained by the ligation reaction is folded to the native tetrameric state. Two subunits of the KcsA tetramer are shown

(Fig. 1a) (1). Central to the function of K+ channels is a structural unit referred to as the selectivity filter (2). In the selectivity filter, a row of four K+ binding sites are created using the backbone carbonyl oxygen atoms from the highly conserved amino acid sequence T-V-G-Y-G and the side chain of the threonine residue (Fig. 1b) (2). Due to the involvement of the protein backbone, SDM is of limited utility in investigating the ion binding sites in the selectivity filter. This limitation motivated the development of the chemical synthesis of the K^+ channel, KcsA. KcsA is a bacterial K^+ channel but its selectivity and permeation properties are similar to those of bigger and more complex K^+ channels (3). The chemical synthesis of the KcsA channel has been used in investigations of the structure and the function of the selectivity filter (4–6). In addition to the KcsA channel, chemical synthesis has so far been reported for the HCV protease cofactor protein NS4A, the influenza M2 channel, and the bacterial mechanosensitive channel MscL (7–9).

In this chapter, we discuss the protocols that we have developed and optimized for the chemical synthesis of the KcsA K^+ channel. We anticipate that the strategies and protocols described in this chapter will find application in the manipulation of other membrane proteins of interest.

1.1 Synthetic Strategy

The KcsA channel is a tetramer of identical subunits that are 160 amino acids long. A truncated form of KcsA lacking the C-terminal 35 amino acids is used for structural studies (10). Due to its shorter size, the truncated form is a simpler synthetic target. In this chapter, we describe the procedures for the chemical synthesis of this truncated form of the KcsA channel.

The synthesis consists of two steps: the assembly of the KcsA polypeptide and the in vitro folding of the KcsA polypeptide to the native tetrameric state (Fig. 1c). The length of the truncated KcsA subunit (125 amino acids) puts it beyond the limits of solid-phase peptide synthesis (SPPS) (11). SPPS is only efficient for the synthesis of peptides ~50–60 amino acids long (11). The KcsA subunit is therefore assembled from two component peptides by native chemical ligation (NCL) (12). In NCL, a peptide with an N-terminal Cys (N-Cys) reacts with an α-thioester peptide to link the two peptides with a native peptide bond at the ligation site. Val70 is used as the ligation site as structural and functional analysis indicates that a Cys substitution at position 70, which is required for the NCL reaction, is well tolerated (5). The synthesis of the KcsA polypeptide therefore requires a peptide thioester corresponding to residues 1–69 (the N-peptide) and an N-Cys peptide corresponding to residues 70–123 (the C-peptide). An advantage of the Val70 ligation site is that the regions of interest in the KcsA channel such as the selectivity filter (residues 75–79) and the cavity region (residues 100–108) are contained within the C-peptide. This allows us to use a semisynthetic approach in which SPPS is used for the C-peptide while the N-peptide thioester is obtained by recombinant means. The advantage of the semisynthetic approach is that it reduces the peptide synthesis steps involved while still permitting the use of SPPS to modify the region of interest ((13), see Note 1)

The amino acid substitution A98→G at the "gating hinge" region allows functional measurements to be easily carried out on the truncated channels (14). The A98G substitution is therefore incorporated into the semisynthetic channels used for functional studies. The amino acid substitutions (Q58A, T61S, R64D), which render the KcsA channel sensitive to block by agitoxin$_2$, are also incorporated into the semisynthetic channel (15). This allows us to use block by AgTx$_2$ to confirm that the electrical activity observed is due to ion movement through the KcsA channel (6).

The second synthesis step is the in vitro folding of the semisynthetic polypeptide to the native state. In vitro folding is necessary because the synthesis steps only provide the unfolded polypeptide, which has to be folded to the native state for structural or functional studies (see Note 2). The KcsA channel is folded in vitro using lipid vesicles (16).

2 Materials

2.1 Synthesis of the C-Peptide

1. Boc-Gly-PAM resin (Advanced Chemtech, Louisville, KY).
2. Boc-protected amino acids.
3. Dimethylformamide (DMF).
4. Dimethylsulfoxide (DMSO).
5. Methylene Chloride (DCM).
6. Diisopropylethylamine (DIE).
7. Trifluoroacetic acid (TFA).
8. 2-(1H-benzotriazol-1-yl)-1,1,3,3-tetramethyluronium hexafluorophosphate (HBTU).
9. Ethanolamine, redistilled.
10. Ninhydrin test reagents: monitor 1, monitor 2, monitor 3, 60% ethanol.
11. p-Cresol.
12. Diethyl ether, anhydrous.
13. 2-Mercaptoethanol (BME).
14. 2,2,2-Trifluoroethanol (TFE).
15. Acetonitrile, HPLC grade.
16. 50-mL glass peptide synthesis vessel (Chemglass, Vineland, NJ).
17. 20-mL glass scintillation vials.
18. Glass stir rods.
19. Nitrogen gas source.
20. Heating block.
21. Anhydrous hydrofluoric acid (HF).
22. Type I HF cleavage apparatus (Peptide Institute, Inc., Osaka, Japan).

23. Buffer A: H_2O, 0.1% TFA.

24. Buffer B: 90% acetonitrile, 10% H_2O, 0.1% TFA.

25. C4 reverse-phase analytical and preparative HPLC columns.

2.2 Recombinant Expression and Purification of the N-Peptide Thioester

1. *Escherichia coli* BL21 competent cells.

2. LB ampicillin plates + 0.1% glucose: 10 g tryptone, 5 g yeast extract, 5 g NaCl, 15 g agar, 1 g glucose, and 100 mg ampicillin per L H_2O.

3. LB broth: 10 g tryptone, 5 g yeast extract, and 10 g NaCl per L H_2O.

4. Dual-fusion protein expression plasmid: KcsA channel residues 1–69 sandwiched between glutathione-*S*-transferase (GST, plasmid vector pGEX 4T-2, GE Healthcare, Piscataway, NJ) at the N-terminus and the *gyr*A intein (plasmid vector pTXB3, New England Biolabs, Ipswich, MA) at the C-terminus.

5. Isopropyl β-D-1-thiogalactopyranoside (IPTG).

6. Cell resuspension buffer: 50 mM Tris–HCl, pH 7.5, 0.2 M NaCl, 1 mM $MgCl_2$, deoxyribonuclease I (DNAse I, 0.01 mg/mL), and 1 mM phenylmethanesulfonyl fluoride (PMSF).

7. Triton X-100.

8. Inclusion body wash buffer: 50 mM Tris–HCl, pH 7.5, 0.2 M NaCl, 2 M urea, and 1% Triton X-100.

9. Pellet solubilization buffer: 50 mM Tris–HCl, pH 7.5, 0.2 M NaCl, 1% *N*-lauryl sarcosine (NLS), 6 M urea, ultra pure grade.

10. Dialysis buffer 1: 50 mM Tris–HCl, pH 7.5, 0.2 M NaCl, 1% Triton X-100.

11. SDS sample buffer: 10% glycerol (v/v), 62.5 mM Tris–HCl pH 6.8, 2% SDS, 0.01 mg/mL bromophenol blue, 5% BME.

12. 2-Mercaptoethanesulfonic acid, sodium salt (MESNA).

13. Chitin beads (New England BioLabs, Ipswich, MA).

14. Thrombin (Roche Diagnostics, Indianapolis, IN)

15. Trichloroacetic acid (TCA).

16. Acetone, 0.1% TFA.

17. 2,2,2-Trifluoroethanol (TFE).

18. Buffer A: H_2O, 0.1% TFA.

19. Buffer B: 90% acetonitrile, 10% H_2O, 0.1% TFA.

20. C4 reverse-phase analytical and preparative HPLC columns.

2.3 Semisynthesis of the KcsA Channel

1. 0.1 M sodium phosphate buffer, pH 7.8.

2. Thiophenol.

3. Dithiothreitol (DTT).

4. Asolectin: L-α-phosphatidylcholine, soy ~20% (Avanti, Alabaster, AL).

5. Asolectin hydration buffer: 50 mM MES, pH 6.4, 0.3 M KCl, and 10 mM DTT.

6. Dialysis buffer 2: 50 mM Tris–HCl, pH 7.5, 0.3 M KCl.

7. Cyclohexane.

8. 2-(N-morpholino) ethanesulfonic acid (MES).

9. Water bath sonicator (Laboratory Supplies Co, Hicksville, NY).

10. Co^{2+} resin (Clontech, Palo Alto, CA).

11. n-Decyl-β-ᴅ-maltopyranoside (DM).

12. Column buffer: 50 mM Tris–HCl, pH 7.5, 0.15 M KCl, 0.25% DM.

13. Imidazole.

14. Superdex S-200 column (GE Biosciences, Piscataway, NJ).

---

3 Methods

3.1 Synthesis of the C-Peptide

1. The C-peptide is assembled by manual solid-phase peptide synthesis using a slightly modified version of in situ neutralization/HBTU activation (17). The sequence of the C-peptide is CE**TATTV**GYGDLYP**VT**LWGRL**VAVVV**M**VGGIT**S FGL**VT**AALATWFVGREQERRG. We carry out the synthesis on a 0.5 mmol scale. Weigh out 2.2 mmol of the Boc-protected amino acids and 0.5 g of the Boc-Gly-PAM resin (1 mmol/g) into glass vials. All amino acids used for peptide synthesis were Boc-protected. For instances where side-chain protection was required, the following amino acid derivatives were used: N-α-t.-Boc-N^G-tosyl-L-arginine, N-α-t.-Boc-L-aspartic acid β-cyclohexyl ester, N-α-t.-Boc-S-methylbenzyl-L-cysteine, N-α-t.-Boc-L-glutamic acid γ-cyclohexyl ester, N-α-t.-Boc-O-benzyl-L-serine, N-α-t.-Boc-O-benzyl-L-threonine, N-α-t.-Boc-N-in-formyl-L-tryptophan, and N-α-t.-Boc-O-2-bromo benzyloxycarbonyl-L-tyrosine.

2. Swell the Boc-Gly-PAM resin in DMF for a minimum of 4 h to overnight, prior to synthesis. Transfer the resin to a peptide synthesis vessel and thoroughly wash with DMF with two 30-s "flow wash" cycles. In the flow wash technique, the resin is washed using a continuous flow of DMF while keeping a constant volume of DMF above the resin bed. In between flow washes, break up the resin and wash the sides of the vessel with DMF.

3. De-protect the Boc-group by treating the resin with TFA. Use two small portions of TFA to initially wash the resin, and then treat the resin with 2–3 volumes of TFA for two 1-min intervals.

Stir the resin during TFA treatment with a glass rod. Flow wash the resin thoroughly with DMF (2×30-s).

4. Activate 2.2 mmol of the first amino acid (Arg) to be coupled by adding 4 mL of 0.5 M HBTU in DMF (2.0 mmol) followed by 1 mL of DIEA (5.74 mmol). Mix the solution by vortexing and allow activation to occur for at least 5 min at room temperature.

5. Couple the first amino acid by adding the pre-activated amino acid solution to the resin and stirring with a gentle flow of nitrogen gas for 15 min. Stir the resin with a glass rod if excessive foaming occurs. Wash the resin thoroughly with DMF (2×30-s). Take a small aliquot of the resin (~5 mg) and perform a ninhydrin reaction to determine the extent of coupling (18). The extent of coupling should be greater than 98%. If the extent of coupling is lower, then repeat the coupling step.

6. Repeat steps 3–5 for all the amino acids in the sequence. β-Branched amino acids (depicted in bold in the sequence) are known to undergo slow and incomplete couplings (11). As a precaution, all β-branched amino acids are double-coupled using 0.5 M HBTU in DMSO. Any non-β-branched amino acids which occur within a stretch of β-branched residues are also double-coupled.

7. After the final de-protection step, neutralize the resin by treating it 2× with 10% DIEA in DMF for 1 min. Wash the resin thoroughly with DMF.

8. Treat the resin 3× with 10% ethanolamine in 5% H_2O/DMF for 30 min to remove the formyl protecting group from the Trp residues. Wash the resin with DMF and then wash thoroughly with DCM before drying overnight in vacuo.

9. Cleavage of the peptide from the resin and global de-protection is carried out using anhydrous HF in the Type I HF-Reaction Apparatus. HF is a highly toxic and corrosive gas. HF cleavage is carried out in specialized equipment that is commercially available. In our laboratory, we use a Type I HF-Reaction Apparatus from Peptide Institute, Inc. (Osaka, Japan).

10. Following HF cleavage, precipitate the peptide by adding 100 mL of cold diethyl ether containing 5% BME. Remove the ether solution by filtration, wash extensively with cold diethyl ether, and dissolve the crude peptide in 50% TFE: Buffer A (see Note 3). Confirm the presence of the desired C-peptide in the crude by RP-HPLC of analysis on a 50–100% B gradient on a C4 analytical column and electrospray mass spectrometry (ES-MS) (Fig. 2a, b). Lyophilize the crude peptide solution.

11. Redissolve the lyophilized peptide crude in a minimal amount of 50% TFE: Buffer A. Purify the peptide with RP-HPLC using

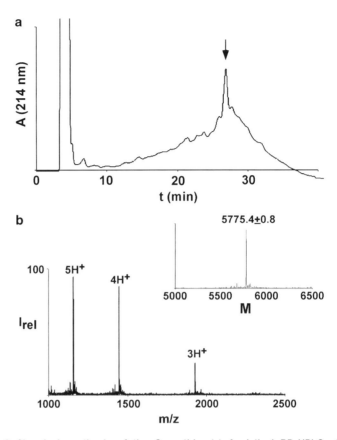

Fig. 2 Chemical synthesis of the C-peptide. (**a**) Analytical RP-HPLC of the C-peptide crude obtained after the HF cleavage reaction (gradient 50–100% Buffer B on an analytical C4 column). The peak corresponding to the C-peptide is marked with an *arrow*. (**b**) ES-MS of the purified C-peptide (*inset*, reconstructed spectrum, expected mass = 5,775.6)

a linear 65–90% Buffer B gradient over 60 min on a C4 preparative column.

12. The fractions containing the pure C-peptide are identified using ES-MS. The pure fractions are pooled and lyophilized to obtain the pure peptide as a white powder. The peptide is stored at −80°C until use.

3.2 Recombinant Expression and Purification of the N-Peptide Thioester

We use a dual-fusion construct for the recombinant expression of the N-peptide thioester (Fig. 3a) (19). The dual-fusion construct consists of KcsA channel residues 1–69 sandwiched between glutathione-*S*-transferase at the N-terminus and the *gyr*A intein at the C-terminus. In the dual-fusion protein, GST directs expression to inclusion bodies while the *gyr*A intein is used to introduce the α-thioester group at the C-terminus of the KcsA N-peptide (see Note 4).

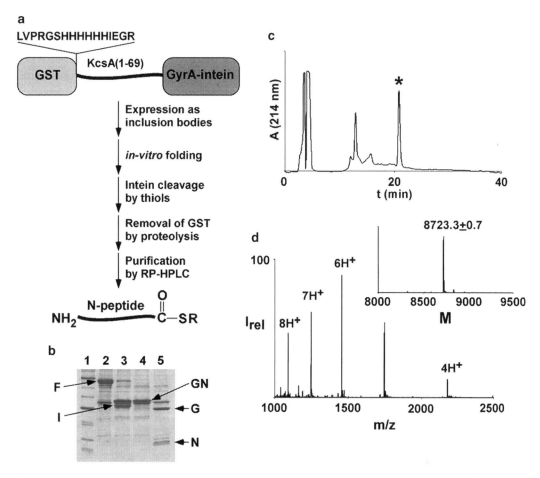

Fig. 3 Recombinant expression and purification of membrane spanning thioester peptides. (**a**) The dual-fusion strategy used for the overexpression and purification of the N-peptide thioester. (**b**) SDS-PAGE gel showing (*lane 1*) the dual-fusion protein (F) after partial purification; (*lane 2*) cleavage of the fusion protein with MESNA to generate the *gyr*A intein (I) and the GST-KcsA N-peptide thioester (GN); (*lane 3*) the flow through from the chitin column contains the GST-N-peptide thioester; (*lane 4*) thrombin proteolysis of GN to separate the GST tag (G) from the N-peptide thioester (N). (**c**) Analytical RP-HPLC of the thrombin cleavage reaction after the TCA precipitation and acetone washes (gradient 50–100% Buffer B on an analytical C4 column). The peak corresponding to the N-peptide thioester is marked with an *asterisk*. (**d**) ES-MS of the N-peptide thioester (*inset*, reconstructed spectrum, expected mass = 8,721.1)

3.2.1 Cell Growth

1. Transform *Escherichia coli* BL21 competent cells with the dual-fusion protein expression plasmid using standard protocols. Select for transformants on LB ampicillin plates containing 0.1% glucose. Incubate the plates overnight at 37°C. Use 1 plate per liter of culture (19).

2. For cell growth, scrape the cells from each plate into 1 L of LB broth containing 0.1 mg/mL of ampicillin. Grow cells at 37°C and monitor cell growth by the optical density (OD) at 600 nm.

At an OD of 1.0, take a 1.0 mL sample of the cell culture (un-induced sample). Induce protein expression by adding IPTG to 1 mM. The IPTG stock is prepared in sterile water.

3. Continue cell growth at 37°C for 3 h. After 3 h of induction, take a 1.0 mL sample of the cell culture (induced sample). The cells are then harvested by centrifugation and the cell pellet is stored at −80°C.

4. Pellet the 1.0 mL un-induced and induced samples, and solubilize the cell pellet by boiling in 0.2 mL of SDS sample buffer. Run the samples on a 12% SDS-PAGE gel to confirm fusion protein expression. Fusion protein expression can be easily detected by comparing the induced sample lane to the un-induced sample lane.

3.2.2 Protein Purification

1. Resuspend cells in the cell resuspension buffer. Lyse cells by sonication (other methods for cell lysis can also be used). Add Triton X-100 to 1% and stir gently at RT for 30 min. Centrifuge the cell lysate at $10,000 \times g$ to separate the soluble and the insoluble fractions.

2. The insoluble fraction contains the inclusion bodies. It is washed 2× with the inclusion body wash buffer. The pellet remaining after the 2 M urea washes is solubilized in the pellet solubilization buffer. The solution is centrifuged at $10,000 \times g$ to remove any insoluble material. Triton X-100 is added to 1% (v/v) and the solution is dialyzed against 20 volumes of dialysis buffer 1. In the dialysis step, the dual-fusion protein folds due to the removal of the denaturants: urea and NLS. The dialysis is carried out overnight at 4°C.

3. Following dialysis, the *gyr*A intein is cleaved by the addition of solid MESNA to 0.1 M. The intein cleavage reaction is carried out at RT with gentle stirring. The progress of the cleavage reaction is monitored by SDS-PAGE (Fig. 3b). The cleavage is generally 80–90% complete after 24 h.

4. The GST-KcsA thioester formed by the MESNA cleavage reaction is separated from the other proteins by passing the cleavage mixture through a chitin column. The chitin column binds the cleaved intein and the dual-fusion protein through the chitin binding domain at the C-terminus. The binding capacity of the chitin resin is 2 mg/mL. We determine the size of the chitin column to be used from a rough estimate of the amount of the cleaved intein and the uncleaved dual-fusion protein, as determined from a Coomassie-stained gel (Fig. 3b).

5. Fractions of the flow-through the chitin column that contain the GST-KcsA thioester are identified by SDS-PAGE and pooled. The KcsA N-peptide thioester is proteolyzed from the GST protein using thrombin. We use 1 U of thrombin per

2 mg of the GST-fusion protein. The cleavage is carried out overnight at room temperature and tested by SDS-PAGE (Fig. 3b). The cleavage is generally complete after overnight incubation.

6. The thrombin cleaved proteins are precipitated by TCA. TCA is added to 15% (w/v) and incubated at 4°C for 30 min. The TCA precipitate is collected by centrifugation at $3,000 \times g$. In addition to the proteins, TCA also precipitates Triton X-100. The TCA precipitate is therefore washed (2×) with ice cold acetone (containing 0.1% TFA) to remove Triton X-100. The precipitate after the acetone washes is dissolved in 50% TFE: Buffer A and the KcsA N-peptide thioester is purified by RP-HPLC (Fig. 3c).

7. RP-HPLC purification is carried out on a C4 column using a 50–100% Buffer B gradient. Fractions containing the purified N-peptide thioester are identified by ES-MS (Fig. 3d). The fractions are pooled and lyophilized. The protocol described yields around 2 mg of the N-peptide thioester per liter of cell culture.

3.3 Semisynthesis of the KcsA Channel

3.3.1 Ligation

1. The ligation reaction between the N-peptide thioester and the C-peptide is carried out in the presence of 1% (w/v) SDS. The detergent, SDS, is used to keep the hydrophobic N- and C-peptides soluble during the ligation reaction. To aid in solubilization, the peptides are co-lyophilized in the presence of SDS (see Note 5). We use a slight excess of the C-peptide (1.5-fold) over the N-peptide in the ligation reaction.

2. The ligation reaction is carried out in 0.1 M sodium phosphate buffer. Prior to the reaction, a small sample is withdrawn and treated with an equal volume of SDS sample buffer to serve as the 0-min control. The ligation reaction is initiated by the addition of thiophenol to 2%. The progress of the reaction is monitored by SDS-PAGE (Fig. 4a). The ligation reactions are usually complete within 12–24 h. The ligation reaction is quenched by the addition of DTT to 0.1 M.

3. The semisynthetic KcsA polypeptide can be purified by RP-HPLC. However, the purification is not efficient and gives very low yields of the purified material. We, therefore, carry out the folding reaction on the ligation reaction mixture without any purification.

3.3.2 In Vitro Folding and Purification of Semisynthetic KcsA

1. In vitro folding of the semisynthetic KcsA polypeptide is accomplished using lipid vesicles. Folding is observed in many types of lipid vesicles. We commonly use soybean lipids (asolectin). For the formation of lipid vesicles, a desired amount of the solid asolectin (200 mg of lipid per milliliter of the ligation reaction)

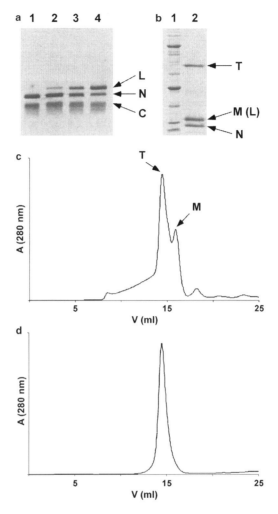

Fig. 4 Ligation and in vitro folding of the semisynthetic KcsA channel. (**a**) SDS-PAGE gel showing the ligation reaction between the N-peptide thioester (N) and the C-peptide (C) to form the KcsA polypeptide (L). (*Lane 1*) 0 min, (*lane 2*) 2 h, (*lane 3*) 6 h, and (*lane 4*) 24 h. (**b**) SDS-PAGE gel showing the semisynthetic KcsA channel after partial purification using metal affinity chromatography. The unfolded monomeric (M, which corresponds to the ligation product), the folded tetrameric KcsA (T), and the unreacted N-peptide are indicated. (**c**) Size-exclusion chromatography is used to separate the tetrameric folded semisynthetic KcsA channels (T) from the unfolded protein (M). (**d**) Size-exclusion chromatography of the purified semisynthetic KcsA channel. In panels (**c**) and (**d**), a Superdex S-200 column and a running buffer of 50 mM Tris–HCl, 150 mM KCl, and 5 mM DM are used

is dissolved in cyclohexane and then lyophilized. Lyophilization converts the asolectin from a granular state to a fine powder. The asolectin powder is hydrated in asolectin hydration buffer at a lipid concentration of 20 mg/mL for 30 min at RT. Small unilamellar vesicles are then formed by sonication in a water bath sonicator.

2. For folding, the ligation mixture is diluted tenfold into asolectin vesicles. The final SDS concentration should be 0.1% or less as higher SDS levels interfere with the folding of the KcsA channel. The extent of folding is monitored by SDS-PAGE. Due to its extreme stability, the KcsA channel migrates as a tetramer on SDS-PAGE and the appearance of a tetramer band indicates the folding of the semisynthetic KcsA channel. We generally allow the folding reaction to continue overnight at RT.

3. The folded semisynthetic KcsA channels are purified by metal affinity chromatography followed by size-exclusion chromatography. Prior to these purification steps, the folding reaction mixture is dialyzed (2×, 12 h each step) against dialysis buffer 2. This dialysis step removes the DTT present in the folding reaction, which is required as DTT interferes with the metal affinity chromatography step. The lipid vesicles after dialysis are solubilized by decylmaltoside (DM, 2% w/v) at RT for 2 h. The solubilized proteins are then bound to Co^{2+} resin. The proteins bound to the Co^{2+} resin are washed with column buffer and then eluted with column buffer containing 0.5 M imidazole. The elution fraction from the Co^{2+} column contains the folded semisynthetic KcsA channels along with the unfolded channels and the unreacted N-peptides (Fig. 4b).

4. We separate the folded semisynthetic channels from the unfolded proteins and the unreacted peptides by size-exclusion chromatography (Fig. 4c, d). Size-exclusion chromatography is carried out on a Superdex S-200 column using column buffer. The fractions containing the folded semisynthetic KcsA channel are identified by SDS-PAGE. These protein fractions are then pooled and concentrated. The concentrated protein is dialyzed against 100 volumes of column buffer, overnight at 4°C.

5. The semisynthetic KcsA channels obtained after dialysis are frozen in liquid nitrogen and stored at –80°C until use. The semisynthetic KcsA channels can be used for structure determination or can be reconstituted into lipid vesicles for functional studies. The reader should refer to publications describing these studies for further details (5, 6, 20).

4   Notes

1. A key factor to consider when designing the synthesis is the length of the protein. The two-part ligation strategy described in this chapter can be used for the synthesis of proteins that are less than 130 amino acids or proteins longer than 130 amino acids but have regions of interest that are within 50–60 amino acids of the N- or the C-termini. This constraint is due to the

limits of SPPS. For proteins longer than 130 amino acids with centrally located regions of interest, a three-part ligation strategy will be required for the assembly of the polypeptide. Protocols for accomplishing a three-part semisynthesis of the KcsA channel have been reported (21).

2. An important consideration before designing the chemical synthesis of a membrane protein is determining whether the protein can be folded in vitro. This is essential as synthetic steps only provide the unfolded polypeptide which has to be folded to the native state for functional characterization and structural investigations. This is a very challenging step and the protocol used depends upon the specific protein. The protocol described in this chapter for the in vitro folding of the KcsA channel has been also been successfully applied for the in vitro folding of the NaK and K$_v$AP ion channels.

3. In our experience a challenging aspect of the SPPS is the solubilization of peptide crude and the purification of the peptide from the crude. A set of solvent mixtures that can assist in solubilization of the peptide crude have been previously reported (19).

4. In the dual-fusion construct used for recombinant expression of the N-peptide thioester, GST is used as the N-terminal fusion protein to direct protein expression to inclusion bodies (19). Other proteins that direct expression to inclusion bodies can also be used for this purpose. Similarly, the choice of the *gyr*A intein for introduction of the thioester at the C-terminus of the N-peptide is based on the efficient refolding of the *gyr*A intein. Any other intein that can be refolded in vitro can be substituted for the *gyr*A intein in the dual-fusion construct.

5. For a successful ligation, the peptide fragments have to be maintained in the soluble state. For membrane spanning polypeptides, this is accomplished by carrying out the ligation reaction in the presence of detergents. The detergent SDS has been used for the ligation reaction in the semisynthesis of the KcsA channels. Another detergent to try is dodecylphosphocholine. The ligation reaction can also be carried out on peptides that have been incorporated into lipid vesicles (21–23).

Acknowledgements

This research was supported by grants to FIV from the NIH (GM087546), a Scientist Development Grant from the American Heart Association (0835166N) and a Pew Scholar Award.

References

1. Hille B (2001) Ion channels of excitable membranes, 3rd edn. Sinauer Associates Inc, Sunderland, MA

2. MacKinnon R (2004) Potassium channels and the atomic basis of selective ion conduction (Nobel lecture). Angew Chem Int Ed 43: 4265–4277

3. LeMasurier M, Heginbotham L, Miller C (2001) KcsA: it's a potassium channel. J Gen Physiol 118:303–314

4. Valiyaveetil FI, Sekedat M, MacKinnon R, Muir TW (2004) Glycine as a D-amino acid surrogate in the K(+)-selectivity filter. Proc Natl Acad Sci U S A 101:17045–17049

5. Valiyaveetil F, Sekedat M, MacKinnon R, Muir T (2006) Structural and functional consequences of an amide-to-ester substitution in the selectivity filter of a potassium channel. J Am Chem Soc 128:11591–11599

6. Valiyaveetil FI, Leonetti M, Muir TW, MacKinnon R (2006) Ion selectivity in a semisynthetic K+ channel locked in the conductive conformation. Science 314:1004–1007

7. Bianchi E, Ingenito R, Simon RJ, Pessi A (1999) Engineering and chemical synthesis of a transmembrane protein: the HCV protease cofactor protein NS4A. J Am Chem Soc 121:7698–7699

8. Kochendoerfer GG, Salom D, Lear JD, Wilk-Orescan R, Kent SBH, DeGrado WF (1999) Total chemical synthesis of the integral membrane protein influenza A virus M2: role of its C-terminal domain in tetramer assembly. Biochemistry 38:11905–11913

9. Clayton D, Shapovalov G, Maurer J, Dougherty D, Lester H, Kochendoerfer G (2004) Total chemical synthesis and electrophysiological characterization of mechanosensitive channels from Escherichia coli and Mycobacterium tuberculosis. Proc Natl Acad Sci U S A 101: 4764–4769

10. Doyle DA, Cabral JM, Pfuetzner RA, Kuo AL, Gulbis JM, Cohen SL, Chait BT, MacKinnon R (1998) The structure of the potassium channel: molecular basis of K+ conduction and selectivity. Science 280:69–77

11. Kent S (1988) Chemical synthesis of peptides and proteins. Annu Rev Biochem 57: 957–989

12. Dawson PE, Muir TW, Clarklewis I, Kent SBH (1994) Synthesis of proteins by native chemical ligation. Science 266:776–779

13. Muir T (2003) Semisynthesis of proteins by expressed protein ligation. Annu Rev Biochem 72:249–289

14. Valiyaveetil FI, Sekedat M, Muir TW, MacKinnon R (2004) Semisynthesis of a functional K+ channel. Angew Chem Int Ed 43: 2504–2507

15. MacKinnon R, Cohen SL, Kuo AL, Lee A, Chait BT (1998) Structural conservation in prokaryotic and eukaryotic potassium channels. Science 280:106–109

16. Valiyaveetil F, Zhou Y, MacKinnon R (2002) Lipids in the structure, folding, and function of the KcsA K+ channel. Biochemistry 41: 10771–10777

17. Schnolzer M, Alewood P, Jones A, Alewood D, Kent SB (1992) In situ neutralization in Boc-chemistry solid phase peptide synthesis. Rapid, high yield assembly of difficult sequences. Int J Pept Protein Res 40:180–193

18. Kaiser E, Colescott RL, Bossinger CD, Cook PI (1970) Color test for detection of free terminal amino groups in the solid-phase synthesis of peptides. Anal Biochem 34:595–598

19. Valiyaveetil F, MacKinnon R, Muir T (2002) Semisynthesis and folding of the potassium channel KcsA. J Am Chem Soc 124:9113–9120

20. Heginbotham L, LeMasurier M, Kolmakova-Partensky L, Miller C (1999) Single streptomyces lividans K(+) channels: functional asymmetries and sidedness of proton activation. J Gen Physiol 114:551–560

21. Komarov AG, Linn KM, Devereaux JJ, Valiyaveetil FI (2009) Modular strategy for the semisynthesis of a K+ channel: investigating interactions of the pore helix. ACS Chem Biol 4:1029 1038

22. Hunter CL, Kochendoerfer GG (2004) Native chemical ligation of hydrophobic (corrected) peptides in lipid bilayer systems. Bioconjug Chem 15:437–440

23. Otaka A, Ueda S, Tomita K, Yano Y, Tamamura H, Matsuzaki K, Fujii N (2004) Facile synthesis of membrane-embedded peptides utilizing lipid bilayer-assisted chemical ligation. Chem Commun (Camb) 15:1722–1723

Chapter 2

Chemical Derivatization and Purification of Peptide-Toxins for Probing Ion Channel Complexes

Zhengmao Hua and William R. Kobertz

Abstract

Ion channels function as multi-protein complexes made up of ion-conducting α-subunits and regulatory β-subunits. To detect, identify, and quantitate the regulatory β-subunits in functioning K$^+$ channel complexes, we have chemically derivatized peptide-toxins that specifically react with strategically placed cysteine residues in the channel complex. Two protein labeling approaches have been developed to derivatize the peptide-toxin, charybdotoxin, with hydrophilic and hydrophobic bismaleimides, and other molecular probes. Using these cysteine-reactive peptide-toxins, we have specifically targeted KCNQ1-KCNE1 K$^+$ channel complexes expressed in both *Xenopus* oocytes and mammalian cells. The modular design of the reagents should permit this approach to be applied to the many ion channel complexes involved in electrical excitability as well as salt and water homoeostasis.

Key words Ion channel, Regulatory subunit, Scorpion toxin, Electrophysiology

1 Introduction

Ion channels co-assemble with different regulatory β-subunits to afford membrane-embedded complexes with diverse ion-conducting and voltage-gating properties that fulfill the ion permeation needs of a wide range of cells and tissues (1–4). Ion channel complex assembly is vital for proper cellular function, as mutations that prevent proper complex formation give rise to neurological, cardiac, and muscular diseases (5–7). While the importance of these membrane-embedded complexes is widely appreciated, it remains challenging to detect, identify, and quantitate the regulatory subunits in functioning ion channel complexes.

To overcome some of these challenges, we have chemically derivatized peptide-toxins to specifically react with target cysteine residues placed in the regulatory β-subunits of functioning K$^+$ channel-KCNE complexes (8, 9). The peptide-toxin is derivatized on the nonbinding side of the toxin with a bismaleimide linker that

Matthew R. Banghart (ed.), *Chemical Neurobiology: Methods and Protocols*, Methods in Molecular Biology, vol. 995,
DOI 10.1007/978-1-62703-345-9_2, © Springer Science+Business Media New York 2013

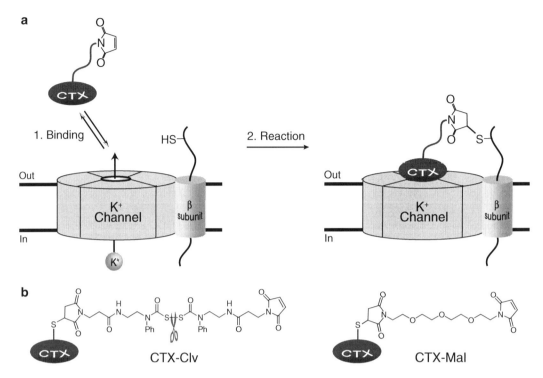

Fig. 1 (**a**) Cartoon depiction of the derivatized peptide-toxin (CTX) binding to the K⁺ conducting subunit followed by the tethering reaction between the maleimide and cysteine residue in the regulatory β-subunit. (**b**) Molecular structures of the derivatized peptide-toxins: CTX-Clv and CTX-Mal. indicate the TCEP cleavable bond in CTX-Clv

converts these reversible inhibitors into cysteine-reactive reagents. Application of a low concentration (~ nM) of the derivatized toxin to cells expressing the K⁺ channel complex results in first binding to the complex (Fig. 1a) followed by a rapid and irreversible reaction with an appropriately positioned cysteine residue in the KCNE regulatory subunit. The specificity and rapid labeling of the regulatory subunit arise from the toxin's avidity for the ion-conducting subunit, which increases the local concentration of the cysteine-modifying group (maleimide) near the target cysteine in the ion channel complex. Without this toxin-aided increase in effective concentration, the bimolecular reaction between the maleimide and cysteine would not measurably occur. Toxin binding and chemical modification are monitored by measuring the current flowing through the K⁺ channel complexes since toxin binding—both reversible and irreversible—completely blocks ion permeation.

The approach has several technical and experimental advantages: (a) Peptide-toxins can be readily expressed as fusion proteins in *E. coli* and the proteolytically cleaved, fully folded toxin can be purified by reverse-phase HPLC. Alternatively, peptide-toxins can be synthesized by solid-phase peptide synthesis, refolded, and similarly purified by HPLC. (b) The folded toxins are very stable and compatible with organic solvents, enabling the chemical derivatization of the toxin with hydrophobic compounds, linkers, and probes. (c) The chemically derivatized toxins are extremely water soluble, reducing the nonspecific binding observed with many hydrophobic reagents and probes. (d) Ion channels where no known toxin exists can be mutated to bind a particular toxin with high affinity (10–14). We have previously exploited this approach to specifically target exogenously expressed KCNQ1 K⁺ channel complexes in *Xenopus* oocytes (8, 9).

The following procedures detail two approaches (Fig. 2) to label the peptide-toxin, charybdotoxin, with a hydrophobic,

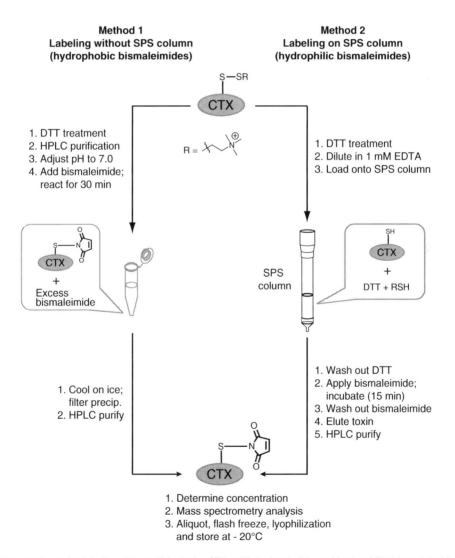

Fig. 2 Two methods for labeling the peptide-toxin, CTX, with hydrophobic and hydrophilic bismaleimides

reductant-cleavable linker (Fig. 1b, CTX-Clv) and with a water-soluble, non-cleavable linker (Fig. 1b, CTX-Mal). The modularity of the approach combined with the assortment of peptide-toxins available to specifically inhibit different classes of ion channels should allow this approach to be readily applied to a variety of membrane-embedded ion channel complexes.

2   Materials

2.1   Bismaleimide Modification of Peptide-Toxins Containing a Modifiable Cysteine Residue

1. Recombinant charybdotoxin, CTX-R19C, was purified as the methanethiosulfonate ethyltrimethylammonium (MTSET)-protected disulfide, as described by Shimony and Miller (15).

2. 1 M DL-dithiothreitol (DTT) dissolved in deionized water and stored in single use (1.5 mL) aliquots at −20°C.

3. Bismaleimides (Pierce).

4. Organic cosolvent for dissolving hydrophobic bismaleimides (dimethylformamide (DMF) and/or acetonitrile (ACN)).

5. 1 M Potassium phosphate, pH 7.1.

2.2   Sulfopropyl-Sephadex (SPS) Separation of Modified Toxins

1. Sulfopropyl-sephadex resin (SPS, dry bead 40–125 μm).

2. 1 mM ethylenediaminetetraacetic acid (EDTA), pH 7.1.

3. SPS Buffer C (low salt): 10 mM KCl, 10 mM potassium phosphate at pH 7.4, 7.1, and 6.0.

4. SPS Buffer D (high salt): 1 M KCl, 10 mM potassium phosphate, pH 6.0.

5. Bio-Rad glass "Econo-Column" column (1×10 cm).

2.3   Reverse-Phase HPLC Purification of Modified Toxins

1. Solvent A (aqueous): 0.1% trifluoroacetic acid (TFA) in deionized water.

2. Solvent B (organic): acetonitrile (ACN, HPLC grade).

3. Analytical C18 HPLC column (protein and peptide C18, 5 μm, 4.6×250 mm).

4. Large-volume injection loop (5 mL).

5. Organic solvent compatible 0.45 μm syringe filters (Life Sciences, HPLC certified).

2.4   Modification of K⁺ Channel Complexes with Maleimido-Toxins

1. Reduced glutathione (GSH).

2. Bovine serum albumin (BSA).

3. Peristaltic pump (optional).

4. External recording solution for either *Xenopus* oocytes or mammalian cells.

3   Methods

3.1   Labeling and Purification of CTX-MTSET with a Hydrophobic Bismaleimide

1. CTX-MTSET (16 nmol) is dissolved in 2 mL of SPS Buffer C at *pH 7.4* (see Note 1).

2. The free thiol of CTX-R19C is generated by reduction with DTT (2 μL of a 1 M stock solution) for 45 min (see Note 2).

3. The reaction mixture containing reduced CTX-R19C is directly injected onto an analytical C18 HPLC column that is pre-equilibrated in Solvent A: 95%; Solvent B: 5% and eluted with a gradient of 5–40% B over 35 min (see Note 3). Exact elution gradient is shown in Fig. 3 (see Note 4). Toxin signal

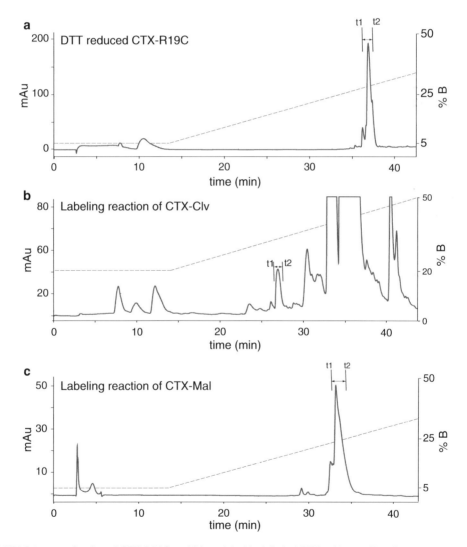

Fig. 3 HPLC traces of reduced CTX-R19C and bismaleimide-labeled CTX-adducts. Absorbance signal is measured at 280 nm (*left axis*). Dashed lines indicate the solvent gradient in %B (*right axis*). The collected peaks are bracketed between t1 and t2. (**a**) DTT-reduced CTX-R19C; (**b**) labeling reaction of CTX-Clv; (**c**) labeling reaction of CTX-Mal

(underivatized and derivatized) is best detected by monitoring at 280 nm.

4. The peak containing the DTT-free, reduced CTX-R19C is collected (Fig. 3a) and the solution is adjusted to pH 7.0 with 1 M potassium phosphate, *pH 7.1* (see Note 5).

5. The solution of neutralized, reduced CTX-R19C is slowly added to a solution of 16 µmol of bismaleimide in 100 µL of an organic solvent (ACN, DMF) with vigorous swirling and allowed to react for 30 min at room temperature (see Notes 6 and 7).

6. The reaction mixture is placed on ice for 10 min to precipitate excess unreacted bismaleimide, which is removed by filtration using an organic solvent-compatible syringe filter (GHP Acrodisc 0.45 µm, Pall Gelman Laboratory).

7. The mono-derivatized CTX-R19C is HPLC-purified using a 20–50% B gradient over 30 min (Fig. 3b).

8. The peak corresponding to the mono-derivatized CTX-R19C (CTX-Clv) is collected (see Note 8), the majority of which is immediately aliquoted into microcentrifuge tubes (30–50 µL aliquots), flash frozen with liquid nitrogen, lyophilized to dryness without heating, and stored at –20°C (see Note 9).

9. A small sample (75–100 µL) is used to both determine the concentration of the aliquots and to determine the molecular weight of the derivatized toxin.

10. To determine the concentration of the aliquots, a quartz UV cell is blanked on the HPLC elution buffer based on the approximate Solvent B % of the collected peak. For example, in Fig. 3c, a mixture of 75% Solvent A and 25% Solvent B was used as the blank. Under these conditions, 1 OD @ A_{280} = 1 mg/mL for CTX (see Note 10). Overall yield of the purified maleimide-toxins is ~20–40%.

11. For mass spectrometry, the collected sample is directly infused into an electrospray ionization (ESI) mass spectrometer in the positive mode (Fig. 4). As expected for a protein with many arginine and lysine residues, multiply charged species are observed in the mass spectrum. Since the toxins are HPLC-purified with 0.1% TFA, TFA-adducts are also observed in the mass spectra (see Note 11).

3.2 Labeling and Purification of CTX-MTSET with a Hydrophilic Bismaleimide

1. CTX-MTSET (10 nmol) in 0.5 mL of SPS Buffer C at *pH 7.4* is reduced by adding 1 µL of 0.5 M DTT (see Note 5).

2. After 30 min, the reaction is diluted with 5 mL of 1 mM EDTA, *pH 7.4*, and is applied to a SPS column (0.3 g) equilibrated in SPS Buffer C.

3. The excess DTT is washed out with 60 mL of SPS Buffer C, *pH 7.1* (slight pressure can be applied with a peristaltic pump or handheld motorized pipetting device).

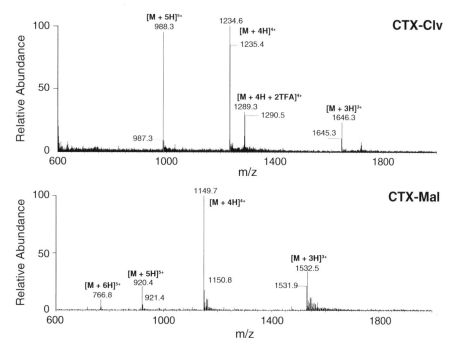

Fig. 4 Electrospray mass spectra of purified CTX-Clv and CTX-Mal in the positive mode. The different charge species for the derivatized toxins are denoted in . Expected MW for CTX-Clv: 4933 (methionine oxidized); **+3H**, 1,646; **+4H**, 1,234; **+5H**, 988. Expected MW for CTX-Mal: 4593 (methionine oxidized); **+3H**, 1,532; **+4H**, 1,149; **+5H**, 920; **+6H**, 767

4. Dissolve bismaleimide (10 μmol) in 1 mL SPS Buffer C, *pH 7.1* (see Note 7). For bismaleimides with intermediate hydrophobicity, ~100 μL of an organic cosolvent can be used to first dissolve the bismaleimide. The bismaleimide solution can contain up to ~10% organic solvent (see Note 6).

5. Slowly apply the 10 mM bismaleimide solution to the top of column and allow the entire solution to penetrate the SPS resin. Cap the bottom of the column (to prevent leakage) and incubate for 15 min.

6. Wash out excess linker with 50 mL of SPS Buffer C, *pH 6.0.* (precious, in-house synthesized or expensive bismaleimides can be partially recovered by collecting the first 20 mL of the eluent and purifying it in 4 mL fractions by reverse-phase HPLC) (see Note 3).

7. Elute the mono-derivatized toxin with SPS Buffer D, *pH 6.0*; collect 4×4 mL fractions.

8. The four fractions are individually desalted and HPLC-purified on an analytical reverse-phase C18 column using the 5–40% B gradient (35 min) in Fig. 3c. Fractions 1 and 2 are usually the most concentrated fractions (see Note 3).

9. The peak corresponding to the mono-derivatized CTX-R19C is collected (see Note 8), aliquoted, quantitated, characterized, and stored as described in Subheading 3.1.

3.3 Application of CTX-Maleimides to Cells Expressing Cysteine-Containing K+ Channel Complexes

1. Mammalian cells or *Xenopus* oocytes expressing ion channel complexes harboring extracellular cysteine residues should be incubated in 2 mM reduced glutathione to prevent cysteine oxidation that cannot be reversed with DTT or TCEP. The media should be changed daily to maintain an extracellular reducing environment (see Note 12). Cells can be incubated without glutathione; however, the labeling efficiency will be compromised.

2. The cells are first treated with 2 mM TCEP (or DTT) in external recording solution for 5 min to fully reduce the extracellular cysteines on the cell surface.

3. The TCEP-treated cells are rinsed and voltage-clamped and the external recording solution should contain 50 μg/mL BSA to prevent the toxins from nonspecifically binding to the recording chamber, tubing, etc.

4a. The CTX-maleimides are dissolved in the appropriate electrophysiological recording solution (containing BSA) and are immediately perfused into the recording chamber. For CTX-sensitive KCNQ1-KCNE1 complexes, we have used 10 nM CTX-maleimide to achieve >90% inhibition. The pH of the recording solution should be between 7.0 and 7.6 for optimal labeling (see Note 5).

4b. For the more hydrophobic CTX-maleimides, the toxin is first dissolved in a minimal amount of 30% ACN in external recording solution and then diluted with external recording solution such that the final toxin solution contains less than 0.3% ACN.

5. If continuous perfusion is required for stable current measurements, a peristaltic pump can be used to recycle the toxin solution back into the recording chamber.

6. Test the maleimido-toxin on the "wild-type" construct to ensure that the toxin does not react with any of the endogenous cysteines in the ion channel complex as in Fig. 5a. If irreversibility is observed, a systematic removal of the cysteines in the external half of the ion channel complex is needed. If complete reversibility is observed, these experiments serve to characterize the derivatized toxin's binding affinity and off- and on-rates to make sure that the chemical tinkering did not significantly perturb toxin binding to the ion channel complex.

7. The cells expressing ion channel complexes with a target cysteine residue are treated with CTX-maleimide for 5 min and the excess reagent is washed out. For optimally placed cysteine

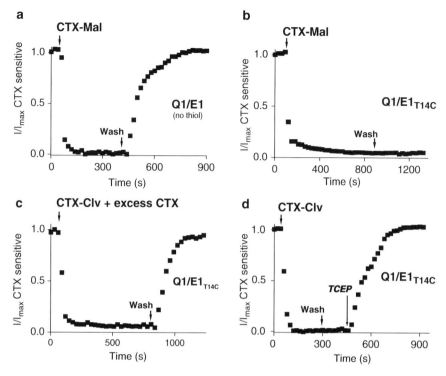

Fig. 5 Utilization of CTX-Mal and CTX-Clv for labeling and identifying KCNE regulatory subunits in functioning KCNQ1-KCNE1 K⁺ channel complexes. (**a**) Q1/E1 complexes lacking extracellular cysteine residues are only reversibly inhibited by 10 nM CTX-Mal, demonstrating the requirement for a target cysteine residue. (**b**) Single cysteine-containing Q1/E1$_{T14C}$ complexes were irreversibly inhibited by 10 nM CTX-Mal and did not reverse upon removal (wash) of excess reagent. (**c**) Irreversible inhibition of Q1/E1$_{T14C}$ by CTX-Clv requires binding to the ion-conducting subunit since excess (500 nM) competitive inhibitor (CTX unlabeled) completely prevents chemical labeling and irreversibility. (**d**) Q1/E1$_{T14C}$ complexes were first irreversibly inhibited by 10 nM CTX-Clv, which was reversed by TCEP cleavage (1 mM) of the linker connecting the maleimide to the toxin

residues, the reaction is usually complete within 100–150 s (see Notes 13 and 14).

8. After the excess toxin is washed out, the extent of labeling is determined by the amount of irreversible inhibition (Fig. 5b).

9. To ensure that the irreversible labeling observed is dependent on toxin binding, perform a competition experiment with unlabeled toxin (Fig. 5c). Successful prevention of labeling with excess competitor demonstrates that the nonspecific, bimolecular reaction between the maleimido-toxin and the ion channel complex is not appreciably occurring.

10. For CTX-maleimides with chemically cleavable linkers, the reagent to cleave the linker (in this case TCEP) is perfused into the bath using the BSA-containing recording solution (Fig. 5d). The cleavage reaction is monitored as a return of current to pretreatment levels.

4   Notes

1. CTX-R19C is purified as the mixed disulfide (CTX-MTSET) to prevent both toxin dimerization (via disulfide bond formation) and cysteine oxidization. In addition, the mixed disulfide is required to cyclize the N-terminus of the toxin (15).

2. DTT reduction of CTX-MTSET should not last longer than 60 min since longer incubation times (as well as DTT concentrations >1 mM) lead to reduction of the internal disulfide bonds of the toxin. Accordingly, the stronger reducing agent, TCEP, should not be used because it rapidly reduces and induces unfolding of the toxin.

3. A 5 mL injection loop is essential for rapid purification of the derivatized toxins due to the milliliter volumes needed for the SPS column separation and chemical reactions. In addition, we never inject more than 4.5 mL since we have noticed that some of the sample exits the loop no matter how slowly/gently the sample is loaded with volumes greater than 4.5 mL.

4. Much of the work of the HPLC is to remove the excess salts from the SPS Buffers or to remove DTT; therefore, our standard gradient involves an initial isocratic segment for 10–15 min (Fig. 3), which removes the majority of the side products.

5. The pH of the solutions for the cysteine modifications and maleimide reactions is critical for obtaining toxins with an intact, reactive maleimides. Thus, we have emboldened the pH of the solutions since it changes during the preparation of the derivatized toxins. In general, thiol-modifying reagents require a pH > 7.0, which generates a significant concentration of the reactive thiolate anion. However, once the reactions are complete, the pH is lowered to ~6.0, which reduces the hydrolysis of the remaining intact maleimide.

6. The organic cosolvents will need to be optimized for each individual bismaleimide. The major complication is due to the bismaleimide precipitating out when the toxin is added to the bismaleimide solution. The reverse addition—adding the bismaleimide in neat organic solvent to the toxin in a water/organic mixture—always resulted in precipitation in our hands.

7. The concentration of the maleimide is critical for efficient labeling. It is preferable to operate at least in the single-digit mM range, ensuring a rapid reaction between thiol and maleimide while minimizing maleimide hydrolysis. Since bismaleimides can potentially react with two CTX-R19C toxins, a 100–1,000-fold molar excess of bismaleimide (compared to toxin) is used to eliminate this potential side reaction.

8. Identifying the maleimido-toxin in the HPLC chromatogram can be challenging, in particular, if Subheading 3.1 is used.

During the development phase, we collect each peak in the chromatogram and identify the peaks using ESI mass spectrometry. The positively charged toxins ionize very well, providing strong, easily identifiable signals, even in the presence of the ion-suppressing TFA.

9. Although we rapidly freeze-dry the HPLC-collected samples, the CTX-maleimides and bismaleimides are reasonably stable (~h) in HPLC solvents that contain TFA.

10. The classic ninhydrin assay (16) can be used as an alternative to the "quick and dirty" UV spec method for determining derivatized toxin concentration. We also always compare the apparent binding affinity of the newly synthesized derivatized toxin to underivatized toxin (see step 6, Subheading 3.3).

11. For CTX-R19C, the lone methionine residue in the toxin is often oxidized, which shifts the molecular weight of the toxin by 16 Da.

12. We have consistently made the glutathione-containing media fresh each day from powder; however, similar to DTT, concentrated glutathione aliquots should be stable at –20°C.

13. The exact location of the target cysteine in the ion channel complex is critical for a rapid and quantitative reaction (> 99%). We have found the rate and the extent of reaction can vary significantly (40–100%) on the location of the target cysteine. We have not tested whether this variation is due to steric hindrance or orientation of the target cysteine. Thus, it is prudent to make several single-cysteine mutations in the region of interest in the ion channel complex.

14. Linker length of the bismaleimide is the second variable that needs to be optimized. We have found that a ~20 Å linker is sufficient for labeling the external face of the pore-forming unit (S5-P-S6) of voltage-gated K$^+$ channels; a ~40 Å linker seems sufficient to reach most extracellular regions of the voltage-sensing unit (S1–S4), though we have been able to label the extended S3–S4 loop of the *Shaker* voltage sensor with the shorter, 20 Å linker.

References

1. McCrossan ZA, Abbott GW (2004) The MinK-related peptides. Neuropharmacology 47: 787–821

2. Lu R, Alioua A, Kumar Y, Eghbali M, Stefani E, Toro L (2006) MaxiK channel partners: physiological impact. J Physiol 570:65–72

3. Hanlon MR, Wallace BA (2002) Structure and function of voltage-dependent ion channel regulatory beta subunits. Biochemistry 41: 2886–2894

4. Dolphin AC (2003) Beta subunits of voltage-gated calcium channels. J Bioenerg Biomembr 35:599–620

5. Choi E, Abbott GW (2007) The MiRP2-Kv3.4 potassium channel: muscling in on Alzheimer's disease. Mol Pharmacol 72:499–501

6. Splawski I, Shen J, Timothy KW, Lehmann MH, Priori S, Robinson JL, Moss AJ, Schwartz PJ, Towbin JA, Vincent GM, Keating MT (2000) Spectrum of mutations in long-QT

syndrome genes. KVLQT1, HERG, SCN5A, KCNE1, and KCNE2. Circulation 102: 1178–1185

7. Abbott GW, Butler MH, Bendahhou S, Dalakas MC, Ptacek LJ, Goldstein SA (2001) MiRP2 forms potassium channels in skeletal muscle with Kv3.4 and is associated with periodic paralysis. Cell 104:217–231

8. Morin TM, Kobertz WR (2008) Counting membrane-embedded KCNE β-subunits in functioning K⁺ channel complexes. Proc Natl Acad Sci U S A 105:1478–1482

9. Morin TM, Kobertz WR (2007) A derivatized scorpion toxin reveals the functional output of heteromeric KCNQ1-KCNE K⁺ channel complexes. ACS Chem Biol 2:469–473

10. Soh H, Goldstein SA (2008) I SA channel complexes include four subunits each of DPP6 and Kv4.2. J Biol Chem 283:15072–15077

11. Chen H, Kim LA, Rajan S, Xu S, Goldstein SA (2003) Charybdotoxin binding in the IKs pore demonstrates two MinK subunits in each channel complex. Neuron 40:15–23

12. Gross A, Abramson T, MacKinnon R (1994) Transfer of the scorpion toxin receptor to an insensitive potassium channel. Neuron 13: 961–966

13. Alabi AA, Bahamonde MI, Jung HJ, Kim JI, Swartz KJ (2007) Portability of paddle motif function and pharmacology in voltage sensors. Nature 450:370–375

14. Bosmans F, Martin-Eauclaire MF, Swartz KJ (2008) Deconstructing voltage sensor function and pharmacology in sodium channels. Nature 456:202–208

15. Shimony E, Sun T, Kolmakova-Partensky L, Miller C (1994) Engineering a uniquely reactive thiol into a cysteine-rich peptide. Protein Eng 7:503–507

16. Rosen H (1957) A modified ninhydrin colorimetric analysis for amino acids. Arch Biochem Biophys 67:10–15

Chapter 3

Using Yeast to Study Potassium Channel Function and Interactions with Small Molecules

Sviatoslav N. Bagriantsev and Daniel L. Minor Jr.

Abstract

Analysis of ion channel mutants is a widely used approach for dissecting ion channel function and for characterizing the mechanisms of action of channel-directed modulators. Expression of functional potassium channels in potassium-uptake-deficient yeast together with genetic selection approaches offers an unbiased, high-throughput, activity-based readout that can rapidly identify large numbers of active ion channel mutants. Because of the assumption-free nature of the method, detailed biophysical analysis of the functional mutants from such selections can provide new and unexpected insights into both ion channel gating and ion channel modulator mechanisms. Here, we present detailed protocols for generation and identification of functional mutations in potassium channels using yeast selections in the potassium-uptake-deficient strain SGY1528. This approach is applicable for the analysis of structure–function relationships of potassium channels from a wide range of sources including viruses, bacteria, plants, and mammals and can be used as a facile way to probe the interactions between ion channels and small-molecule modulators.

Key words SGY1528, Potassium channel, Ion channel, Yeast, Gating

---

1 Introduction

Accumulation of high concentrations of intracellular potassium is essential for life. In *Saccharomyces cerevisiae*, two potassium transporters Trk1p and Trk2p serve as the major components of the potassium-uptake system (1, 2) and allow the yeast to grow under low (0.5–1.0 mM) potassium conditions. The double knockout of these transporters, *trk1Δtrk2Δ*, fails to survive when subjected to low potassium conditions but can be rescued by the heterologous expression of a number of different types of functional potassium channels from viruses (3–5), archaea (6), bacteria (7–9), plants (10–16), and mammals (7, 17–23, 30). This phenotypic rescue is driven by the fact that, unlike animal cells, the membrane potential of yeast is exceptionally negative due to the action of the plasma membrane proton ATPase (24). This situation is enhanced in *trk1Δtrk2Δ* yeast (25), which have an estimated membrane potential of

Matthew R. Banghart (ed.), *Chemical Neurobiology: Methods and Protocols*, Methods in Molecular Biology, vol. 995,
DOI 10.1007/978-1-62703-345-9_3, © Springer Science+Business Media New York 2013

~–300 mV (24). Thus, in the presence of a functional, open potassium channel, this negative potential is able to drive intracellular potassium accumulation from the surroundings through the overexpressed channel and allow for rescue of growth. When combined with random or directed mutagenesis protocols, this system can serve as a powerful selection method to identify novel and functional ion channel mutations that provide the starting point for dissection of structure–function relationships. It should be noted that to date, the channels that have been used successfully in this system all have the property of being open under hyperpolarized conditions. For a broader review on different approaches to study ion channel structure and function see ref. 26.

In this chapter, we provide a detailed protocol for generating a random mutagenic library and selection for gain-of-function mutations of a potassium channel in the *trk1Δtrk2Δ* yeast strain SGY1528 (21) based on a number of studies in which we have used the system to explore both viral and mammalian potassium channel function (4, 5, 18, 19). Additionally, we describe a basic approach towards using the yeast strain to screen for modulator-resistant or modulator-sensitive mutations.

2 Materials

The protocol described here assumes the knowledge of standard molecular biology techniques, *E. coli* manipulations, and media recipes. For general manipulations with yeast see ref. 27. Use at least ≥98.5% purity-grade chemicals and Milli-Q-grade water.

2.1 Equipment

1. Tools for colony picking and streaking: while sterile standard pipette tips or microbiological loops can be used for picking and streaking yeast colonies, we routinely use flat toothpicks sterilized in dry conditions in scintillation vials.

2. Replica plating tool and sterile velveteen squares (Scienceware).

3. Microtube mixer Tomy MT-360 (Tommytech), or a similar device.

4. Plasmid DNA isolation kit (Qiagen), or similar.

5. DNA gel extraction kit (Qiagen), or similar.

6. 0.5 mm acid-washed glass beads, ≥100 μl per sample.

7. Sterile Whatman paper filter discs (5–10 mm diameter).

8. 0.22 μm sterilization filter units (e.g., Nalgene, cat# 127-0020 or similar).

2.2 Yeast and Plasmids

The protocol describes procedures optimized for the derivative of the *Saccharomyces cerevisiae* strain W303, SGY1528 (MAT*a*; *ade2-1*; *can1-100*; *his3-11,15*; *leu2-3,112*; *trp1-1*; *ura3-1*; *trk1::HIS3*;

trk2::TRP1) (21). To express potassium channels in yeast, we have been routinely using the multicopy pYes2 (2 μ, *URA3*, AmpR) shuttle vector (Invitrogen), modified by replacing the galactose-inducible GAL4 promoter with methionine-repressible MET25 promoter (28) as described elsewhere (19). This change allows for an efficient way of controlling channel expression by simply adding or removing methionine from the growth medium. The channel of interest is cloned 5′–3′ between *Hin*dIII and *Xho*I sites of the vector.

2.3 Yeast Media Components and Transformation Solutions

1. -Ura/-Met dropout powder: 6.0 g glutamic acid, 2.5 g adenine hemisulfate, 1.2 g arginine, 6.0 g aspartic acid, 1.8 g lysine, 3.0 g phenylalanine, 22.5 g serine, 12.0 g threonine, 2.4 g tryptophane, 1.8 g tyrosine, 9.0 g valine, 1.2 g histidine, 3.4 g leucine. Mix well, and store at room temperature for up to 2 years in moisture-free conditions. This component supplies nucleobases and amino acids that are necessary for yeast growth, except for uracil (for plasmid selection) and methionine (to drive protein expression from the MET25 promoter).

2. 1,000× Vitamin stock solution: 2 mg biotin, 400 ng D-pantothenic acid, 400 mg pyridoxine, 400 mg thiamin, 2 g inositol. Dissolve the components in 1 L water, sterilize through a 0.22 μm filter, and store at −20°C for 2–3 years.

3. 1,000× Trace mineral solution: 50 mg boric acid, 4 mg $CuSO_4$, 10 mg KI, 50 mg $FeCl_3$, 40 mg $MnSO_4$, 90 mg molybdic acid, 40 mg $ZnSO_4$. Dissolve the components in 100 ml water, add 1 ml concentrated HCl, sterilize through a 0.22 μm filter, and store at −20°C for up to 3 years.

4. Li–TE (lithium–Tris–EDTA): 100 mM LiCl, 10 mM Tris–HCl pH 7.5, 1 mM EDTA. Sterilize through a 0.22 μm filter, store at room temperature.

5. PEG–TE (polyethylene glycol–Tris–EDTA): 50% (w/v) PEG-3350 (polyethylene glycol with average molecular weight 3,350), 10 mM Tris HCl pH 7.5, 1 mM EDTA. Sterilize through a 0.22 μm filter, store at room temperature.

2.4 Yeast Media

1. YPAD/100 K, nonselective medium (liquid): 10 g yeast extract, 20 g dextrose, 20 g peptone, 24 mg adenine hemisulfate, 7.46 g KCl (see Note 1). Dissolve the components in 1 L water, sterilize through a 0.22 μm, and store at room temperature for up to 6 months.

2. YPAD/100 K, nonselective medium (plates): 10 g yeast extract, 20 g dextrose, 20 g peptone, 20 g Bacto agar, 24 mg adenine hemisulfate, 7.46 g KCl (see Note 1). Mix the components in 1 L water, sterilize by autoclaving, and pour the plates (25–30 ml per plate). Leave the plates at room temperature for 24 h to dry. Store the plates in a plastic bag at 4°C for up to 6 months.

3. -Ura/-Met 100 K, plasmid-selective synthetic medium (liquid): 1.5 g -Ura/-Met dropout powder, 6.7 g yeast nitrogen base (without amino acids), 20 g dextrose, 7.46 g KCl. Dissolve the components in 1 L water, and bring pH to 6.5 with 1 M Tris (free base) solution. Sterilize the medium through a 0.22 μm filter, and store at room temperature for up to 6 months.

4. -Ura/-Met 100 K, plasmid-selective synthetic medium (plates): 20 g Bacto agar in 500 ml water prepared separately from 1.5 g -Ura/-Met dropout powder, 6.7 g yeast nitrogen base (without amino acids), 7.46 g KCl, water to 450 ml, to pH 6.5 with 1 M Tris (free base) solution. Autoclave both solutions and combine within 10–15 min after autoclaving to avoid premature solidification of the components. Add 50 ml 40% dextrose (sterilized though a 0.22 μm filter) and pour the plates (25–30 ml per plate). Store the plates in a plastic bag at 4°C for up to 1 year.

5. -Ura/-Met 1 K and -Ura/-Met 0.5 K test medium (plates): 15 g LE agarose in 500 mg water prepared separately from 2.1 g arginine (free base), 1.5 g dropout powder -Ura/-Met, 10 g dextrose, 1 ml 1,000× trace minerals stock solution, 1 ml 1,000× vitamin stock solution, 1 ml 1 M $MgSO_4$, 0.1 ml 1 M $CaCl_2$, 334 μl or 167 μl 3 M KCl for -Ura/-Met 1 K and -Ura/-Met 0.5 K medium, respectively, water to 500 ml, pH 6.0 with phosphoric acid and sterilized though a 0.22 μm filter. Autoclave the agar and add the solution to autoclaved agar while hot, and pour plates (25–30 ml per plate). Store the plates in a plastic bag for up to 1 year.

2.5 Molecular Biology Solutions

1. 10× Error-prone Taq buffer: 100 mM Tris–HCl pH 8.3, 500 mM KCl, 70 mM $MgCl_2$, 5 mM $MnCl_2$, 0.1% gelatin.

2. 10× Error-prone dNTP stock solution: 5 mM dATP, 5 mM dGTP, 25 mM dCTP, 25 mM dTTP in Milli-Q-grade water. Aliquot by 15–20 μl and store at –20°C for up to 1 year. Avoid freeze–thaw cycles.

3   Methods

3.1 Preparing Yeast Stock for Long-Term Storage

1. To prepare SGY1528 yeast stock for long-term storage, streak the cells onto YPAD/100 K plates and grow at 30°C until pinhead-sized individual colonies appear (see Note 2).

2. Pick 3–5 healthy colonies, inoculate 5 ml YPAD/100 K liquid, and grow in a ventilated tube overnight at 30°C at constant shaking at ~225 rpm (rotation per minute).

3. Collect the cells by centrifugation for 10 min at $4,000 \times g$; discard supernatant.

4. Resuspend the cells in 2.5 ml sterile 30% glycerol.

5. Aliquot in 0.5 ml samples using screw-cap or regular Eppendorf tubes.

6. Flash-freeze in liquid nitrogen and store at −80°C for up to 3 years.

3.2 Library Generation Using Taq Polymerase

For an excellent collection of various methods for mutagenic library construction, see ref. 29. Here, we present a Taq polymerase-based protocol to generate random library of a full-length gene. To increase the rate of erroneous nucleotide incorporation, PCR is done under error-prone conditions. Of note, the mutations that appear during the initial PCR cycles will be overrepresented in the pool of the mutated PCR products at the end of the reaction. To account for this, we run at least four independent identical PCRs, and combine the PCR products. This protocol generates 1–5 nucleotide substitutions per 1 kb of gene length. The library is generated in a PCR using 40–45 nucleotide long primers that match the regions immediately flanking the targeted gene. Incorporation of the PCR product into the plasmid occurs by homologous recombination between the sites flanking the targeted gene on the PCR product, and the linearized vector backbone (Fig. 1).

1. Clone the channel of interest into a yeast expression vector using the following scheme: A-*restr1-K+CHANNEL-restr2*-B, where A and B are regions on the vector flanking the gene of interest (*K+CHANNEL*), and *restr1* and *restr2* are sites for endonucleases (in case of the modified pYes2 vector, *restr1* and *restr2* are *Hin*dIII and *Xho*I, respectively).

2. To prepare linearized empty vector, cut 10 μg of the construct with *resrt1* and *restr2* endonucleases, resolve by electrophoresis in agarose, isolate the empty vector from the gel in 50 μl Milli-Q water using the QIAquick gel extraction kit (or similar), and measure DNA concentration by measuring optical density at 260 nm (A_{260} of 1.0 = 50.0 μg DNA). Typical concentration of the linearized vector DNA is ~100 μg/μl. It is not necessary to dephosphorylate the vector.

3. Generate mutagenic library in at least four separate identical PCRs (50 μl each). For each reaction, mix the following components in a thin-walled PCR tube:

Template (uncut plasmid carrying the cassette)—20–25 ng.

10× Error-prone buffer—5 μl.

Error-prone dNTP stock solution—4 μl.

Forward primer (A) and reverse primer (B)—2.5 μl of a 20 pmol/μl stock solution (in water) of each primer.

Taq polymerase—5 units.

Water up to 100 μl.

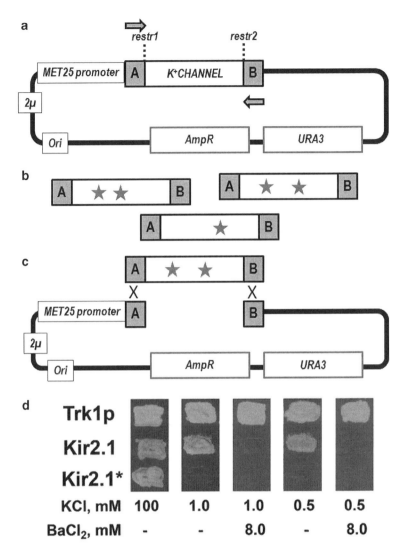

Fig. 1 Library generation and functional expression of a potassium channel in yeast. (**a**) A scheme of the multicopy pYES2-based vector carrying yeast and bacterial origins of replication (2 μ and Ori, respectively), yeast and bacterial selection markers (AmpR and URA3, respectively), and a potassium channel gene (K+ CHANNEL) under the control of the methionine-repressible MET25 promoter. *Arrows* depict forward and reverse primers matching the A and B regions, respectively. (**b**) A library of channel genes carrying random mutations (*stars*) is generated in an error-prone PCR using the plasmid and primers shown in (**a**). (**c**) Diagram depicting heterologous recombination of one channel gene with the linearized vector following preparation by treatment with *restr1* and *restr2* endonucleases. (**d**) Exemplar images of yeast expressing the Trk1, Kir2.1 or the nonfunctional Kir2.1* mutant on medium supplemented with different amounts of KCl or $BaCl_2$. Trk1p and Kir2.1 but not the inactive Kir2.1* mutant enable yeast growth on medium supplemented with 1.0 and 0.5 mM KCl (-Ura/-Met 1 K and -Ura/-Met 0.5 K media in the protocol). Ba^{2+} inhibition of Kir2.1, but not the Ba^{2+}-resistant Trk1p channel correlates with yeast growth on the potassium-depleted test medium

4. Perform PCR cycling using a standard Taq protocol.

   Example

   (a) 2 min at 95°C (initial denaturation).

   (b) 30 s at 95°C (denaturation).

   (c) 30 s at 62°C (annealing).

   (d) [1 min × 1 kb of target DNA length] at 72°C (extension).

   (e) Return to step b, 29 times.

   (f) 5 min at 72°C (final extension).

5. Analyze the PCR by electrophoresis in DNA-grade agarose.

6. Isolate the PCR products from the gel (e.g., using QIAquick gel extraction kit from Qiagen) in 30 µl Milli-Q.

7. Combine the PCR products from the four different reactions in one tube and measure DNA concentration. Typical concentration of the target DNA is ~100 µg/µl. The library is now ready to be mixed with the linearized vector and used for yeast transformation.

3.3 Library Transformation

1. Streak out yeast from a –80°C stock onto a YPAD/100 mM KCl plate, and grow 3–4 days at 30°C.

2. Pick 3–4 medium-sized colonies, inoculate each in 5 ml liquid YPAD/100 mM KCl medium, and grow overnight (14–16 h) in a 50 ml conical tube at 30°C at constant shaking at 225 rpm. To provide adequate ventilation, place the tube cap on and secure it with a tape instead of screwing it on.

3. Add 1.25 ml of the overnight culture to a 250 ml autoclaved flask with 50 ml YPAD/100 mM KCl.

4. Grow at 30°C at constant shaking at 225 rpm until the culture reaches OD600 0.5–0.8 (usually takes 4.5–5 h).

5. Collect the cells by centrifugation in a 50 ml conical tube for 10 min at 4,000 × g, and discard supernatant.

6. Resuspend the cells in 30 ml Li–TE and incubate at 30°C for 1 h without shaking.

7. Collect the cells by centrifugation and resuspend in 850 µl Li–TE (final volume).

8. *For transformation with a plasmid*: transfer 100 µl of the cellular suspension into an Eppendorf tube containing 1–2 µg plasmid DNA and 100 µg of single-stranded carrier DNA preheated for 5 min at 95°C (e.g., herring testes carrier DNA from Clontech, cat#S0277).

   (a) Add 200 µl PEG–TE, mix by gently pipetting the solution up and down 4–5 times, and incubate at room temperature for 1 h.

   (b) Heat-shock for 10 min at 42°C in a heat block.

(c) Incubate at room temperature for 5 min.

(d) Plate 100 µl of each reaction onto one -Ura/-Met 100 K plate, and incubate upside down at 30°C for 3–5 days.

9. *For transformation with a library.* Prepare library mix by combining 3 µg of non-mutagenized vector with 3 µg of the mutagenized gene, and 300 µg of preheated carrier DNA (see above). Combine the DNA mix with 300 µl of yeast cells.

(a) Add two volumes of PEG–TE to one volume of yeast/DNA suspension, and incubate 20 min at 30°C.

(b) Heat-shock for 12 min at 42°C.

10. Plate the whole suspension onto -Ura/-Met 100 K plates (100–150 µl per plate), and grow at 30°C for 3–5 days. If the number of transformants on -Ura/-Met 100 K is low (i.e., less than 50 colonies), increase the amount of cells and/or library DNA. Alternatively, increase the ratio of mutagenized gene to linearized vector.

3.4 Functional Selection for Yeast Expressing Active Potassium Channels

1. Allow 3–5 days for colony formation on -Ura/-Met 100 K medium (see Note 3). Only yeast containing a plasmid expressing the *URA3* gene will grow. Although the absence of methionine in the medium will drive the expression of the potassium channel from the plasmid, the high concentration of potassium in the medium will enable growth regardless of the activity of the channel.

2. Transfer the colonies from -Ura/-Met 100 K onto -Ura/-Met 1 K medium by replica plating, and incubate at 30°C for 2–4 days. Low potassium concentration in the medium will select for cells that express an active potassium channel, such as the yeast Trk1p or mammalian Kir2.1. However, cells expressing poorly functional or nonfunctional potassium channels may grow on this medium too. This may happen due to the presence of residual potassium carried over from the potassium-rich -Ura/-Met 100 K medium. Colonies displaying weak growth due to potassium carryover have a different appearance than those relying on the expressed channel for survival. Comparison with the controls can show some differences. Doubts can be removed by a second replica-plating step onto either -Ura/-Met 1 K or -Ura/-Met 0.5 K.

3. Transfer colonies from -Ura/-Met 1 K onto -Ura/-Met 0.5 K medium by replica plating, and incubate for 3–7 days. Only colonies expressing active potassium channels will grow on this medium.

4. Using a sterile toothpick, scoop up as many colonies as possible from the -Ura/-Met 0.5 K plate and inoculate each into 5 ml of liquid -Ura/-Met 100 K medium (see Note 4) in a 50 ml

conical tube with a loose cap as described above. Pick and inoculate as many colonies as possible, and grow the cultures for 48 h at 30°C at constant shaking at ~225 rpm.

3.5 Plasmid Isolation Using the QIAquick Miniprep Kit, and Retransformation

1. Pellet the yeast by centrifugation for 10 min at $5,000 \times g$ and discard supernatant. Wash the cells by resuspending in water and pelleting. Resuspend the cells in 250 μl of RNAse-containing resuspension buffer P1.

2. Transfer the suspension in an Eppendorf tube containing 100 μl glass beads.

3. Vortex the tube for 4 min at room temperature.

4. Add 350 μl of the lysis buffer P2, and incubate 10 min at room temperature.

5. Add 500 μl of N3, mix well, and centrifuge for 10 min at $15,000 \times g$ at 4°C.

6. Isolate plasmid DNA from the supernatant using the QIAprep columns and the protocol as for a regular DNA miniprep. Briefly, bind DNA to a QIAprep spin column, wash with 750 μl of ethanol-containing wash buffer PE, dry the membrane by additional centrifugation at high speed, and elute DNA in 30 μl water.

7. Use 10–15 μl of the yeast plasmid miniprep to transform *E. coli*, using chemical or electroporation approaches. Regardless of the choice of the transformation method, it is advisable to use bacteria with highest competence rate possible.

8. Prepare minipreps from the bacterial colonies, and transform back into yeast as described above.

9. Pick 1–2 independent yeast colonies for each of the recovered plasmids and patch (~0.5 × 0.5 mm) onto -Ura/-Met 100 K medium. Pass the patch consecutively onto -Ura/-Met 1 K and -Ura/-Met 0.5 K media to confirm the phenotype exhibited by the plasmid in the initial screen (Fig. 1d).

3.6 Yeast Cultivation in the Presence of Chemical Compounds

Sometimes it is desirable to assess activity of a chemical compound against a potassium channel expressed in yeast. In this case, a blocker or an activator of the channel may be directly added to the -Ura/-Met 0.5 K test medium. An activator is expected to facilitate growth, whereas an inhibitor to suppress yeast growth on this medium. Although the compound may be added directly to the hot medium before pouring the plates, we find it more practical to use one of the following two approaches as they avoid potential inactivation of the compound by the hot media. A liquid compound may be applied to a sterile Whatman paper disc (4–5 mm in diameter), and placed onto solidified plates following replica plating of an evenly spread layer of yeast from the -Ura/-Met

100 K or -Ura/-Met 1 K media. In this format, it is possible to test several concentrations of a compound on a single plate, by placing the discs in different sectors of the plate. The halo of yeast growth (or lack thereof) around the disc will be indicative of the effect of the compound. For informative illustrations, see refs. 5, 18. Alternatively, solutions containing the modifier can be applied directly to the plate (e.g., 200 µl of the modifier solution per 10 cm plate with 20–25 ml of solid medium) and spread evenly using a sterile spreader, followed by yeast transfer onto the plate by replica plating. If the plate is prepared properly, the agar will readily imbibe the 200 µl of the modifier solution, and the plate will be ready for replica plating. If the plate stays wet, incubate the plate at room temperature until the modifier solution completely disappears from the surface of the plate.

3.7. Screening for Modulator-Resistant Clones or Modulator-Sensitive Mutations in Potassium Channels

The SGY1528 yeast offers a unique opportunity to pinpoint key structural elements of a potassium channel that mediate its interaction with pharmacological agents. This can be done by utilizing a combination of random or targeted mutagenesis of the potassium channel gene followed by functional selection. For example, it may be desirable to find point mutations in a potassium channel that confer resistance to an inhibitor. In this case, yeast transformed with a library of mutagenized potassium channel are selected on -Ura/-Met 0.5 K medium in the presence of the compound. Yeast cells that bear a mutation in the channel that confers resistance to the inhibitor without compromising channel function will grow on this medium. Plasmids bearing the mutated channels are isolated and retested as described above. To perform functional selection on the -Ura/-Met 0.5 K medium, it is most convenient to apply the compound directly on the plate and spread it evenly using a sterile spreader (see notes in the previous section), and then transfer the yeast onto the plate by replica plating from the -Ura/-Met 100 K or -Ura/-Met 1 K media. Alternatively, the compound may be added onto a Whatman filter and placed in the middle of the -Ura/-Met 0.5 K plate following yeast transfer. For example, see refs. 5, 18.

4 Notes

1. While we routinely use YPAD/100 K medium for nonselective cultivation of the SGY1528 strain, a simpler YPD medium, which is widely used for cultivation of the standard laboratory yeast strains (e.g., W303, 74D), can be used for this purpose. YPD has the same recipe as YPAD/100 K but lacks the additional adenine hemisulfate and KCl. The absence of adenine in the YPD medium will cause the cells to develop a red-colored pigment, but will not impact viability.

2. The strain is extremely sensitive to temperature changes. While *S. cerevisiae* normally tolerates prolonged incubation at 4°C, and is able to grow for several days at 37°C, the SGY1528 strain quickly looses fitness even after an overnight incubation at 4°C, while growth at 37°C is significantly impaired.

3. At this point, it is advisable to pick 50–100 colonies, isolate the plasmids, and sequence in order to determine the rate of mutations in the library.

4. Beware of spontaneous mutants that grow in most stringent potassium-depleted conditions. These colonies are usually characterized by robust growth, and it is therefore advisable not to pick them. If the colony is picked for plasmid isolation and retransformation, the phenotype will not be confirmed in the rescreening round.

Acknowledgements

This work was supported by grants to D.L.M. from NIH (NS065448, MH093603) and the American Heart Association (0740019 N), and to S.N.B. from the Life Sciences Research Foundation. D.L.M. is an AHA established investigator. S.N.B. is a Genentech fellow of the Life Sciences Research Foundation.

References

1. Ko CH, Gaber RF (1991) TRK1 and TRK2 encode structurally related K⁺ transporters. Mol Cell Biol 11:4266–4273

2. Ko CH, Buckley AM, Gaber RF (1990) TRK2 is required for low affinity K+ transport in *Saccharomyces cerevisiae*. Genetics 125:305–312

3. Gazzarrini S, Kang M, Abenavoli A, Romani G, Olivari C, Gaslini D, Ferrara G, van Etten JL, Kreim M, Kast SM, Thiel G, Moroni A (2009) Chlorella virus ATCV 1 encodes a functional potassium channel of 82 amino acids. Biochem J 420:295–303

4. Balss J, Papatheodorou P, Mehmel M, Baumeister D, Hertel B, Delaroque N, Chatelain FC, Minor DL Jr, Van Etten JL, Rassow J, Moroni A, Thiel G (2008) Transmembrane domain length of viral K+ channels is a signal for mitochondria targeting. Proc Natl Acad Sci U S A 105:12313–12318

5. Chatelain FC, Gazzarrini S, Fujiwara Y, Arrigoni C, Domigan C, Ferrara G, Pantoja C, Thiel G, Moroni A, Minor DL Jr (2009) Selection of inhibitor-resistant viral potassium channels identifies a selectivity filter site that affects barium and amantadine block. PLoS One 4:e7496

6. Sesti F, Rajan S, Gonzalez-Colaso R, Nikolaeva N, Goldstein SA (2003) Hyperpolarization moves S4 sensors inward to open MVP, a methanococcal voltage-gated potassium channel. Nat Neurosci 6:353–361

7. Paynter JJ, Andres-Enguix I, Fowler PW, Tottey S, Cheng W, Enkvetchakul D, Bavro VN, Kusakabe Y, Sansom MS, Robinson NJ, Nichols CG, Tucker SJ (2010) Functional complementation and genetic deletion studies of KirBac channels: activatory mutations highlight gating-sensitive domains. J Biol Chem 285:40754–40761

8. Paynter JJ, Sarkies P, Andres-Enguix I, Tucker SJ (2008) Genetic selection of activatory mutations in KcsA. Channels (Austin) 2:413–418

9. Paynter JJ, Shang L, Bollepalli MK, Baukrowitz T, Tucker SJ (2010) Random mutagenesis screening indicates the absence of a separate H(+)-sensor in the pH-sensitive Kir channels. Channels (Austin) 4(5):390–397

10. Anderson JA, Huprikar SS, Kochian LV, Lucas WJ, Gaber RF (1992) Functional expression of a probably *Arabidopsis thaliana* potassium channel in *Saccharomyces cerevisiae*. Proc Natl Acad Sci U S A 89:3736–3740

11. Sentenac H, Bonneaud N, Minet M, Lacroute F, Salmon J-M, Gaymard F, Grignon C (1992) Cloning and expression in yeast of a plant potassium ion transport system. Science 256: 663–665

12. Schachtman DP, Schroeder JI, Lucas WJ, Anderson JA, Gaber RF (1992) Expression of an inward-rectifying potassium channel by the *Arabidopsis KAT1* cDNA. Science 258: 1654–1658

13. Lai HC, Grabe M, Jan YN, Jan LY (2005) The S4 voltage sensor packs against the pore domain in the KAT1 voltage-gated potassium channel. Neuron 47:395–406

14. Nakamura RL, Anderson JA, Gaber RF (1997) Determination of key structural requirements of a K⁺ channel pore. J Biol Chem 272: 1011–1018

15. Rubio F, Schwarz M, Gassmann W, Schroeder JI (1999) Genetic selection of mutations in the high affinity K+ transporter HKT1 that define functions of a loop site for reduced Na+ permeability and increased Na+ tolerance. J Biol Chem 274:6839–6847

16. Rubio F, Gassmann W, Schroeder JI (1995) Sodium-driven potassium uptake by the plant potassium transporter HKT1 and mutations conferring salt tolerance. Science 270: 1660–1663

17. Bichet D, Lin YF, Ibarra CA, Huang CS, Yi BA, Jan YN, Jan LY (2004) Evolving potassium channels by means of yeast selection reveals structural elements important for selectivity. Proc Natl Acad Sci U S A 101: 4441–4446

18. Chatelain FC, Alagem N, Xu Q, Pancaroglu R, Reuveny E, Minor DL Jr (2005) The pore helix dipole has a minor role in inward rectifier channel function. Neuron 47:833–843

19. Minor DL Jr, Masseling SJ, Jan YN, Jan LY (1999) Transmembrane structure of an inwardly rectifying potassium channel. Cell 96: 879–891

20. Sadja R, Smadja K, Alagem N, Reuveny E (2001) Coupling Gbetagamma-dependent activation to channel opening via pore elements in inwardly rectifying potassium channels. Neuron 29:669–680

21. Tang W, Ruknudin A, Yang W, Shaw S, Knickerbocker A, Kurtz S (1995) Functional expression of a vertebrate inwardly rectifying K⁺ channel in yeast. Mol Biol Cell 6:1231–1240

22. Yi BA, Lin YF, Jan YN, Jan LY (2001) Yeast screen for constitutively active mutant G protein-activated potassium channels. Neuron 29:657–667

23. Zaks-Makhina E, Kim Y, Aizenman E, Levitan ES (2004) Novel neuroprotective K+ channel inhibitor identified by high-throughput screening in yeast. Mol Pharmacol 65:214–219

24. Rodriguez-Navarro A (2000) Potassium transport in fungi and plants. Biochim Biophys Acta 1469:1–30

25. Madrid R, Gomez MJ, Ramos J, Rodriguez-Navarro A (1998) Ectopic potassium uptake in trk1 trk2 mutants of *Saccharomyces cerevisiae* correlates with a highly hyperpolarized membrane potential. J Biol Chem 273:14838–14844

26. Minor DL Jr (2009) Searching for interesting channels: pairing selection and molecular evolution methods to study ion channel structure and function. Mol Biosyst 5:802–810

27. Sherman F (2002) Getting started with yeast. Methods Enzymol 350:3–41

28. Kerjan P, Cherest H, Surdin-Kerjan Y (1986) Nucleotide sequence of the *Saccharomyces cerevisiae* MET25 gene. Nucleic Acids Res 14: 7861–7871

29. Arnold FH, Georgiou G (2003) Directed evolution library creations—Methods and protocols. Humana, Totowa, NJ

30. Bagriantsev S, Clark KA, Peyronnet R, Honoré E, and Minor DL Jr (2011) Multiple modalities act through a common gate to control K_{2P} channel function. The EMBO Journal 30: 3594–3606

Chapter 4

A FLIPR Assay for Evaluating Agonists and Antagonists of GPCR Heterodimers

Jessica H. Harvey, Richard M. van Rijn, and Jennifer L. Whistler

Abstract

Calcium signaling plays a major role in the function of cells. Measurement of intracellular calcium mobilization is a robust assay that can be performed in a high-throughput manner to study the effect of compounds on potential drug targets. Pharmaceutical companies frequently use calcium signaling assays to screen compound libraries on G-protein-coupled receptors (GPCRs). In this chapter we describe the application of FLIPR technology to the evaluation of GPCR-induced calcium mobilization. We also include the implications of GPCR hetero-oligomerization and the identification of heteromeric receptors as novel drug targets on high-throughput calcium screening.

Key words Calcium mobilization, G-protein-coupled receptor, GPCR, High-throughput screening, HTS, Heteromer, Oligomerization, Bivalent ligands

1 Introduction

The high-throughput screening (HTS) of large libraries of drug-like compounds is a key aspect of drug discovery. G-protein-coupled receptors (GPCRs) form one of the largest families of drug targets to which HTS is regularly applied (1). GPCRs predominantly exert their function by coupling to heterotrimeric G proteins. Depending on the G_α subunit ($G_{\alpha i}$, $G_{\alpha s}$, $G_{\alpha q}$, $G_{\alpha 12}$), the receptor will signal via adenylyl cyclase/cAMP, phospholipase C/Ca^{2+}, or Rho (2). The activity of receptors that signal through $G_{\alpha q}$ to release Ca^{2+} can be easily monitored in a HTS environment using the fluorescence-based FLIPR calcium assay (Molecular Devices, Sunnyvale, CA). The existence of promiscuous G proteins ($G_{\alpha 16}$) and chimeric G proteins ($G_{\alpha qs}$, $G_{\alpha qi}$) that can provide calcium readouts for GPCRs that would otherwise not signal via the $G_{\alpha q}$ pathway (Fig. 1) adds to the versatility of the FLIPR assay (3, 4).

Traditionally, GPCRs were believed to function as monomers (a single receptor binding one G protein), but, over the past

Matthew R. Banghart (ed.), *Chemical Neurobiology: Methods and Protocols*, Methods in Molecular Biology, vol. 995, DOI 10.1007/978-1-62703-345-9_4, © Springer Science+Business Media New York 2013

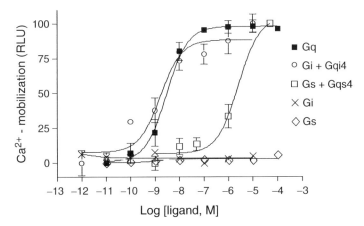

Fig. 1 Measuring Ca²⁺ mobilization by activation of a G_q-, G_i-, or G_s-coupled receptor. HEK-293 cells were transfected with the G_q-coupled cholecystokinin CCK2 receptor, the G_i-coupled mu-opioid receptor (with or without the chimeric G_{qi4}-protein) or the G_s-coupled dopamine D_1 receptor (with or without the chimeric G_{qs4}-protein). Calcium mobilization was measured using the FLIPR assay kit, after activation of the receptors by an appropriate agonist (CCK-8S, etorphine, and dopamine, respectively)

decade, a new consensus has emerged in light of growing evidence that GPCRs can form oligomeric structures. GPCR oligomers can result from the association of one type of receptor with itself (homomers) or from interactions among different receptors (heteromers) (5). Current data suggests that receptor heteromers can have unique pharmacological properties (6, 7) and may be viable drug targets. Thus far, several ligands have been designed or discovered that can bind heteromers. These ligands fall into two groups: monovalent and bivalent (two monophores connected by a linker of appropriate length) ligands. For example, 6′-GNTI is a monovalent ligand that shows potent activity only at the delta–kappa-opioid receptor heteromer (8), whereas the MDAN series are bivalent ligands consisting of a mu-opioid receptor agonist and a delta-opioid receptor antagonist linked by spacers of different lengths (9), some of which produce a lower degree of dependence than morphine or other monovalent ligands that act on mu-opioid receptors alone (Fig. 2).

This chapter describes the application of FLIPR to both conventional as well as bivalent ligands.

2   Materials

2.1   Cell Culture

1. Human embryonic kidney cells (HEK-293) (ATCC, Manassas, VA).

2. Dulbecco's modified Eagle's medium (DMEM) supplemented with 10% fetal bovine serum (GIBCO, Invitrogen, Carlsbad, CA).

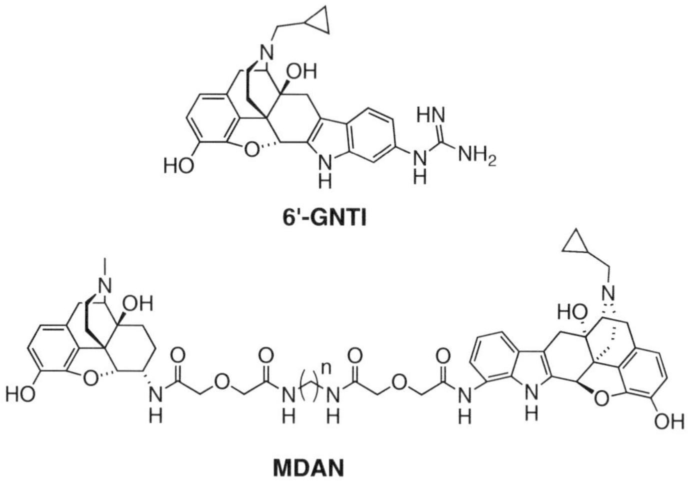

Fig. 2 Heteromer ligands. Monovalent ligand 6′-GNTI and the MDAN bivalent ligand series ($n=5$, MDAN-19)

3. Solution of trypsin (0.25%) and ethylenediamine tetraacetic acid (EDTA).

4. Phosphate buffered saline solution (PBS).

5. Probenecid.

6. Sodium hydroxide.

7. 10× Hank's balanced salt solution (HBSS).

8. 1 M HEPES, pH 7.2–7.5.

9. Poly-D-Lysine.

10. Clear bottom 96-well black plates (Costar, Corning, NY).

11. FLIPR calcium assay kit (Molecular Devices).

12. FlexStation pipette tips (96, clear) (Molecular Devices).

2.2 Transfection of HEK-293 Cells with GPCR and Chimeric G Proteins

1. 1 μg/μl plasmid DNA carrying GPCR of choice (e.g., pcDNA3.1-DOR).

2. 1 μg/μl Gqi5-myr (see Note 1).

3. Lipofectamine (Invitrogen).

4. Opti-MEM (GIBCO, Invitrogen).

3 Methods

3.1 Growing and Seeding Cells

1. HEK-293 expressing the receptor(s) of choice are grown to at least 80% confluent in 10 ml DMEM/10% FBS in a 75 cm^2 cell culture flask in an incubator at 37°C/5% CO$_2$.

2. Count the cells before seeding into a 96-well clear bottom black plate (see Note 2).

3. Dilute cells to 700,000 cells/ml in DMEM/10% FBS and resuspend by gently pipetting up and down (see Note 3).

4. Transfer 100 µl of the diluted cell suspension to each well in the 96-well plate (see Note 4).

5. Incubate the cells for one day in the CO_2-incubator at 37°C/5% CO_2.

3.2 Transfection of Cells with GPCR and/or Chimeric G Protein (Optional, See Note 5)

1. Mix 100 ng of DNA with 50 µl (2 µg/ml) Opti-MEM per well.

2. Mix 0.5 µl Lipofectamine with 50 µl (10 µg/ml) Opti-MEM per well and incubate for 5 min at room temperature (see Note 6).

3. Add the DNA solution to the Lipofectamine solution and incubate for 20 min at room temperature.

4. Aspirate the medium from the cells in the 96-well plate (see Note 7).

5. Gently add 100 µl Opti-MEM to each well using a multichannel pipette.

6. Add 100 µl of the DNA/Lipofectamine solution to each well (final volume is 200 µl) and incubate for 4–5 h in the CO_2-incubator at 37°C/5% CO_2.

7. Aspirate the DNA/Lipofectamine solution, replace it with 100 µl DMEM/10%FBS, and incubate for one day in the CO_2-incubator at 37°C/5% CO_2 (see Note 8).

3.3 Calcium Mobilization Measurement Using the FlexStation

1. Add 10 ml assay buffer to one vial of FLIPR reagent per 96-well plate and mix by vortexing. (see Note 9).

2. Add 100 µl/well FLIPR reagent to the cells in the 96-well plate and incubate for 1 h in a CO_2-incubator at 37°C /5% CO_2 (see Note 10).

3. Prepare the ligands (see Subheading 3.4 below).

4. Optionally, add antagonist to the cells at the appropriate time before measurement (see Note 11).

5. Turn on the FlexStation about 15 min prior to measurement to set up the assay protocol and allow the FlexStation to reach the proper temperature. Select the correct compound plate type, assay plate type, which wells to measure, and the speed and depth at which agonist will be transferred (see Note 12).

6. Set up the protocol to add agonist 10–20 s after the start of the measurement and measure for 2 min with 1.5 s intervals (79 data points). Wavelengths are set at 482 nm for excitation and 525 for emission.

7. Place ligands, cells, and pipette tips in the FlexStation and start measurement (see Note 13).

3.4 Ligand Preparation

3.4.1 Agonist

1. The ligand plate can be prepared after the calcium dye has been added to the cells, during the 60 min waiting period (see Subheading 3.3).

2. Each concentration in the ligand dilution series needs to be prepared at a 5× concentration to account for the dilution that occurs upon mixing with the cells. The dilution series is prepared in the calcium assay buffer (see Note 14).

3. The dilution series is prepared in triplicate on a 96-well plate. The top row can be reserved for solvent controls (for example, buffer only or 0.1% DMSO, see Note 15). Seven concentrations of four different ligands can then be accommodated (alternatively, eight concentrations of three of the ligands can be used if the entire top row is not needed for controls, but three wells are always reserved for the buffer only control).

4. Prepare a solution of the highest concentration of ligand. For a standard log step dilution series pipette, for example, 200 μl of solution into well A1. Fill wells A2–A8 with 180 μl of assay buffer (see Note 16). To prepare the dilution series take 20 μl out of A1 and transfer to A2, mix well, and take 20 μl from A2 and transfer to A3 and so forth. To include 0.5 log steps, prepare a 5× and a 1.67× concentration for the highest concentration of ligand. Transfer this solution to the 5× stock in well A1 and the 1.67× in well A2. Prepare tenfold dilutions A1 to A3 to A5, A2 to A4 to A6, etc.

5. After all ligands have been prepared, and just prior to insertion in the FlexStation, briefly place the plate on a shaker to eliminate air bubbles and ensure mixing.

3.4.2 Antagonist

1. Calcium mobilization does not provide a direct measure of antagonism or inverse agonism. However, antagonism can be determined indirectly by measuring inhibition of agonist-induced calcium release. The antagonist can be added either prior to measurement (10–15 min beforehand), or a solution containing both the agonist and antagonist can be prepared for simultaneous addition (see Note 17) (Fig. 3).

   When multiple dose–response curves are produced in the presence of an increasing concentration of antagonist, the potency of the antagonist can be measured in the form of a pA_2 value obtained from a Schild plot.

3.4.3 Bivalent Ligands

1. Both biochemical (e.g., cross-linking and immunoprecipitation studies) and resonance energy transfer (e.g., bioluminescence resonance energy transfer, fluorescence resonance energy transfer) techniques have shown that receptor oligomerization occurs for many receptors in heterologous expression systems. Additionally, current research implicates receptor heteromers

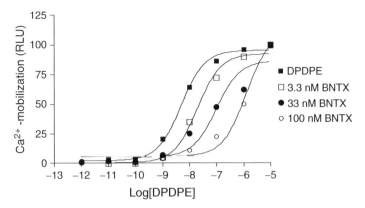

Fig. 3 Measuring antagonism using FLIPR. Increasing doses of DOR antagonist BNTX decrease the potency of DOR agonist DPDPE. Antagonist was incubated with cells for 10 min prior to assay

as potential drug targets, and thus a HTS assay to detect heteromer-selective ligands will be needed. To screen for heteromer-selective ligands, the ligand needs to be tested on three different cell lines: cells expressing receptor 1, cells expressing receptor 2, and cells co-expressing receptor 1 and 2. Using the calcium assay we have been able to identify 6′-GNTI as a potent kappa–delta-opioid heteromer-selective agonist (8). The potency (and thus, the functional activity) of this ligand in cells co-expressing KOR and DOR is distinct from that in cells expressing only KOR or DOR. 6′-GNTI is a weak partial agonist at kappa-opioid receptors (and, thus, functional antagonist in the presence of full agonists such as U50,488), and a potent antagonist at DORs (Fig. 4). On the other hand, we predict MDAN-19 (9), a bivalent ligand comprised of a MOR agonist and a DOR antagonist, to be a potent agonist in cells expressing MOR and a potent antagonist in cells expressing DOR. However, in cells expressing MOR and DOR, it could be either an agonist, or antagonist, or partial agonist, depending on which receptor response predominates in the context of a heteromer.

2. The G protein to which a heteromer couples may be different than the G protein associated with the monomeric receptor(s). Initially, it may not be possible to predict through which G protein the heteromer signals. Thus, the assay can be performed both with and without transfection of a known chimeric G protein. This assay can therefore be used to determine whether heteromerization alters specificity of G protein coupling. For example, neither dopamine D1 nor dopamine D2 receptors couple to G_q when expressed alone (they are $G_{s/olf}$ and $G_{i/o}$ coupled receptors, respectively). However, when these two receptors are co-expressed (and presumably generate heteromers), a dopamine-induced calcium signal is produced,

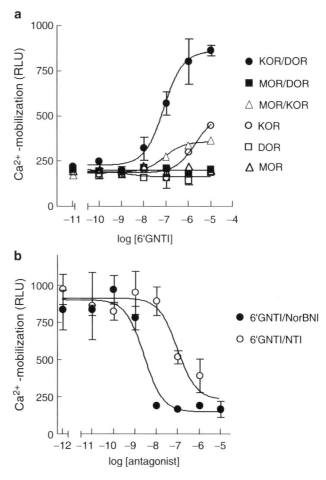

Fig. 4 (a) 6′-GNTI-induced Ca^{2+} release in HEK-293 cells expressing one or two opioid receptor types. Agonist-mediated intracellular Ca^{2+} release was measured in cells expressing the chimeric G protein $\Delta 6$-$G_{qi4\text{-}myr}$ (200 ng for every 40,000 cells). (Reproduced from ref. 8 with permission. Copyright 2005 National Academy of Sciences, USA.) **(b)** Effects of receptor-type-selective antagonists on 6′-GNTI-induced Ca^{2+} release in cells expressing KOR/DOR heteromers. Cells were preincubated for 30 min with increasing doses of NTI or NorBNI and stimulated with 100 nM 6′-GNTI (Reproduced from ref. 8 with permission Copyright 2005 National Academy of Sciences, USA.)

implying that this heteromer couples to G_q. In addition, it is possible that various ligands at a single homomeric or heteromeric receptor could bias signaling to one G protein versus another. Using a combination of chimeric G proteins, one could determine the ligand selectivity for G protein coupling.

3.5 Results Analysis

1. The assay records, from the designated wells, a pre-assay endpoint (to determine basal fluorescence), emitted fluorescence by column, and then a post-assay endpoint. For data analysis,

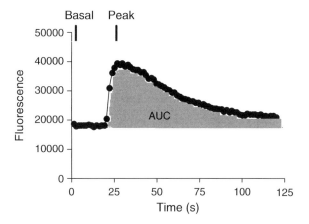

Fig. 5 Agonist-induced calcium mobilization: The calcium wave. Calcium mobilization can be represented as [(peak − basal)/basal] × 1,000 or as area under the curve (AUC)

it is especially important to record the pre-assay endpoint (see Note 18).

2. Data can be saved as SoftMax profile (.pda) or exported as text file, which can be opened in a spreadsheet program (e.g., Microsoft Excel, Open Office spreadsheet).

3. The SoftMax pro datasheet provides the pre- and post-assay endpoints, each time point, and a normalized value of the peak value (peak − basal).

4. Data can be represented as [(peak − basal)/basal] × 1,000 or as area under the curve (AUC, Fig. 5). The pre-assay endpoint is the basal value, while the set of numbers just after the last column analysis and just before the post-assay endpoint are the peak values. The maximum peak height, the AUC, and the shape of the calcium response may all be informative (see Note 19).

4   Notes

1. A different chimeric G protein (3) or no G protein at all can be transfected here, depending on the type of receptor used and the output to be investigated (for example, if a heteromeric receptor is postulated to have a novel G_q-protein coupling different from its monomers, then no G protein may be transfected).

2. Depending on the cell type, cells can be counted using either a hematocytometer or a machine such as a NucleoCounter (New Brunswick Scientific, Edison, NJ).

3. This concentration gives 70,000 cells/well. The FLIPR protocol recommends seeding 50,000 cells/well, but we adjust this number depending on cell type and incubation timeframe.

4. The 96-well plate can be pretreated with poly-d-lysine (PDL) to prevent cells from detaching easily during handling. PDL pretreatment becomes especially important when a relatively harsh method of transfection is used (e.g., cells tolerate FuGENE 6 (Roche) better than Lipofectamine). Additionally, cells should preferably be transferred quickly using a multi-channel pipette or a multidrop dispenser to maintain a homogeneous solution. Otherwise, resuspend the cell solution every couple of minutes.

5. Preferably a cell line stably expressing the receptor of choice (and chimeric G protein) should be used. This will reduce the variation in expression levels that arises from transient transfections. This variation becomes a bigger issue when co-expressing two receptors and looking for novel functions on putative heteromers. In this case at least three cell lines should be generated: one expressing receptor A alone, one expressing receptor B alone, and one expressing both receptor A and receptor B at the same levels as in the homomer cell lines, so that a direct comparison of potency and efficacy can be made (for example, if homomer line A has 50 fmol/mg of receptor A but the heteromer line has 200 fmol/mg of receptor A, both efficacy and potency may be altered, since receptors are competing for transfected G protein as well as endogenous G protein). However, generating heteromer cell lines with expression levels matched to existing homomer lines may not always result in an optimal ratio of heteromer formation. Thus, it is advisable to generate multiple heteromer cell lines with varying ratios of homomers and heteromers, and then select matching homomer cell lines. Therefore, to insure success with this assay, it is best to establish and fully characterize homomer and heteromer cell lines prior to beginning a screen for novel activity on heteromers. The ratio of homomers and heteromers can be determined by several methods including serial co-immunoprecipitation or FRET/BRET analysis.

6. Other transfection methods include FuGENE 6 (Roche), which is gentler to the cells, or calcium phosphate, which is relatively inexpensive and works well with HEK-293 cells.

7. To speed up aspiration a multichannel aspirator head can be used (e.g., multiwell plate washer/dispenser manifold, Sigma).

8. At this stage pertussis toxin (PTX) or cholera toxin (CTX) can be added to the medium. PTX catalyzes the ADP-ribosylation of the α-subunits of the heterotrimeric G proteins G_i, G_o, and G_t, preventing the G proteins from interacting with the receptor. Similarly CTX permanently ribosylates the G_s alpha subunit. The use of the PTX and/or CTX may reveal that a GPCR or heteromer can couple to multiple G proteins, or increase signal output by forcing a GPCR to couple to, for example, a chimeric PTX insensitive G protein (10).

9. The FLIPR assay kit comes equipped with assay buffer. However, the amount provided in the kit may not be sufficient if multiple agonist and antagonist dilution series are prepared. We make additional buffer as follows: For 100 ml, use 10 ml of 10× HBSS, 2 ml 1 M HEPES, 1 ml 0.25 M probenecid [made fresh and dissolved in 1 M NaOH], and 87 ml dH$_2$O).

10. To avoid dislodging cells from the 96-well plate, pipette gently at a 45° angle and avoid touching the bottom of the wells.

11. Since the calcium release occurs in a matter of seconds, the agonist should always be added by the FlexStation.

12. The depth of the pipette tip, the speed, and the volume of transfer can all have an impact on how well the cells remain fixed to the bottom of the well. For a volume of 200 μl a pipette height of 225 μl and a transfer rate of 3 can be used.

13. When placing the ligand plate and assay plate (cell plate) in the FlexStation, make sure they are correctly oriented (i.e., A1 for both plates is in the upper left corner).

14. In order to conserve ligand, all volumes can be reduced by half, and the ligands prepared at a 3× concentration (the assay is then run using 50 μl cells, 50 μl ligand, and 50 μl assay/fluorophore buffer). In practice, this requires 100 μl of ligand per well (The machine may not reliably transfer the desired volume if the well is filled with less than 100 μl). The cells are transfected in the same manner (i.e., placed in 100 μl medium), but before addition of the FLIPR reagent to the cells, 50 μl of buffer is removed from each well of the cell plate. Then 50 μl of FLIPR reagent is added to the cells to give a 100 μl solution to which the 50 μl of 3× ligand will be added. Make sure to change the "compound transfer" conditions under setup to 100 μl initial volume and 125 μl pipette height.

15. The solvent may have an impact on the readout. As seen from Fig. 6, concentrations of 5% DMSO, ethanol, and methanol should be avoided. However, we found that mildly acidic (pH 3, 1 mM HCl) or alkaline solutions (pH 11, 1 mM NaOH) had no significant effect on basal Ca^{2+} mobilization.

16. While 50 μl of ligand/buffer will be transferred from the reagent plate to the assay plate, depending on the plate used (flat bottom, round bottom, v-bottom), in order to make sure the FlexStation can access the solution, more than 50 μl ligand solution will be needed in each well. In general, try to prepare at least 100 μl volume for each concentration.

17. When pre-adding the antagonist, add to each column in timed intervals (e.g., 2 min, using a multichannel pipette) in order to keep the actual preincubation time in line with the time it takes to measure one column in the FlexStation. If the antagonist is added separately, make sure to adjust the concentration of

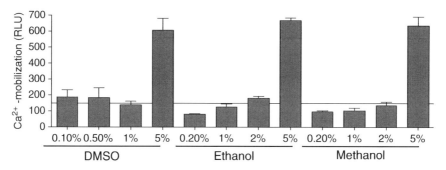

Fig. 6 Effect of DMSO, ethanol, and methanol concentration on calcium mobilization in the FLIPR assay. Ca^{2+} mobilization was measured in HEK-293 cells using the FLIPR assay kit. Cells were stimulated with increasing concentrations of DMSO, ethanol, or methanol to a final concentration of 5% (v/v). The *solid line* represents response to buffer

both the antagonist as well as the agonist to the final volume during measurement: 100 μl medium, 100 μl dye, 50 μl antagonist, and 50 μl agonist require the antagonist and agonist concentrations to be six times more concentrated. Alternatively 25 μl of antagonist and agonist can be used (10× more concentrated).

18. The "pre-assay endpoint" window must be selected when the assay has begun in order for this endpoint to be recorded. Additionally, the empty data file must be saved (as well as the protocol) both before and after the assay. Otherwise the FlexStation may not measure the pre-assay endpoint.

19. The temporal and spatial activity of Ca^{2+} is also known as a Ca^{2+}-wave. The Ca^{2+}-wave relies on release of Ca^{2+} from intracellular stores (via activation of IP_3 and ryanodine receptors) or the influx of Ca^{2+} across the plasma membrane through Ca^{2+} channels. To only study Ca^{2+} mobilization from intracellular stores, cells can be grown in Ca^{2+}-free medium or in the presence of Ca^{2+} chelators such as EGTA. On the other hand, the intracellular Ca^{2+} stores in the endoplasmatic reticulum can be depleted by administration of thapsigargin. Similarly, dantrolene and ryanodine can block ryanodine receptors and limit the passage of Ca^{2+} from smooth ER. The origin of Ca^{2+} release and propagation of a Ca^{2+}-wave may be of interest when studying GPCRs involved in, for example, neurotransmitter release.

Acknowledgements

The authors would like to thank Laura Milan-Lobo for contributions to the dopamine receptor data and Maria Waldhoer for the 6'-GNTI study. This work was funded by the Department of Defense grant DAMD62-10-5-071, National Institute on Drug

Abuse grants R01 DA015232 and DA019958, and funds provided by the State of California for medical research on alcohol and substance abuse through the University of California, San Francisco.

References

1. Eglen RM, Bosse R, Reisine T (2007) Emerging concepts of guanine nucleotide-binding protein-coupled receptor (GPCR) function and implications for high throughput screening. Assay Drug Dev Technol 5: 425–451
2. Marinissen MJ, Gutkind JS (2001) G-protein-coupled receptors and signaling networks: emerging paradigms. Trends Pharmacol Sci 22:368–376
3. Kostenis E, Waelbroeck M, Milligan G (2005) Techniques: promiscuous G alpha proteins in basic research and drug discovery. Trends Pharmacol Sci 26:595–602
4. Conklin BR, Farfel Z, Lustig KD, Julius D, Bourne HR (1993) Substitution of 3 amino-acids switches receptor specificity of G(Q)Alpha to that of G(I)Alpha. Nature 363:274–276
5. Panetta R, Greenwood MT (2008) Physiological relevance of GPCR oligomerization and its impact on drug discovery. Drug Discov Today 13:1059–1066
6. Rios CD, Jordan BA, Gomes I, Devi LA (2001) G-protein-coupled receptor dimerization: modulation of receptor function. Pharmacol Ther 92:71–87
7. Fuxe K, Marcellino D, Guidolin D, Woods AS, Agnati LF (2008) Heterodimers and receptor mosaics of different types of G-protein-coupled receptors. Physiology 23:322–332
8. Waldhoer M, Fong J, Jones RM, Lunzer MM, Sharma SK, Kostenis E, Portoghese PS, Whistler JL (2005) A heterodimer-selective agonist shows in vivo relevance of G protein-coupled receptor dimers. Proc Natl Acad Sci U S A 102:9050–9055
9. Daniels DJ, Lenard NR, Etienne CL, Law PY, Roerig SC, Portoghese PS (2005) Opioid-induced tolerance and dependence in mice is modulated by the distance between pharmacophores in a bivalent ligand series. Proc Natl Acad Sci U S A 102:19208–19213
10. Joshi SA, Fan KP, Ho VW, Wong YH (1998) Chimeric Galpha(q) mutants harboring the last five carboxy-terminal residues of Galpha(i2) or Galpha(o) are resistant to pertussis toxin-catalyzed ADP-ribosylation. FEBS Lett 441: 67–70

Part II

Photochemical Control of Protein and Cellular Function

Chapter 5

Characterizing Caged Molecules Through Flash Photolysis and Transient Absorption Spectroscopy

Joseph P.Y. Kao and Sukumaran Muralidharan

Abstract

Caged molecules are photosensitive molecules with latent biological activity. Upon exposure to light, they are rapidly transformed into bioactive molecules such as neurotransmitters or second messengers. They are thus valuable tools for using light to manipulate biology with exceptional spatial and temporal resolution. Since the temporal performance of the caged molecule depends critically on the rate at which bioactive molecules are generated by light, it is important to characterize the kinetics of the photorelease process. This is accomplished by initiating the photoreaction with a very brief but intense pulse of light (i.e., flash photolysis) and monitoring the course of the ensuing reactions through various means, the most common of which is absorption spectroscopy. Practical guidelines for performing flash photolysis and transient absorption spectroscopy are described in this chapter.

Key words Caged molecule, Uncaging, Photorelease, Kinetics, Photochemistry

1 Introduction

A caged molecule is a biologically inert but photosensitive molecule that has latent biological activity. Light absorption transforms the caged molecule, leading to "photorelease" of a fully bioactive molecule, which can immediately act on biological effectors such as receptors or enzymes. Photochemical reactions are typically very fast, and light can be focused to a spot whose diameter is diffraction-limited (typically <0.5 μm); therefore, focal photolysis of caged molecules is a means to manipulate cell biology and physiology with excellent spatial and temporal precision. Because speed is a chief advantage of using caged molecules, it is important to characterize the kinetics of photorelease.

When a caged molecule absorbs light, it can become activated and transform itself into a product molecule. The photochemical reaction is most simply represented by reaction Scheme 1:

$$C + Photon \rightarrow I \rightarrow P, \qquad (Scheme\ 1)$$

Matthew R. Banghart (ed.), *Chemical Neurobiology: Methods and Protocols*, Methods in Molecular Biology, vol. 995, DOI 10.1007/978-1-62703-345-9_5, © Springer Science+Business Media New York 2013

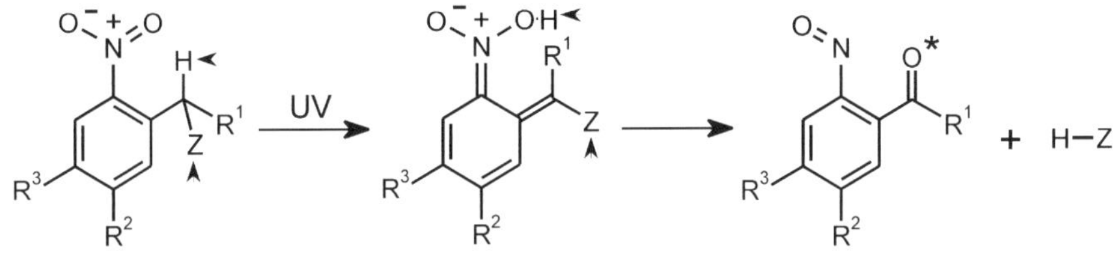

Fig. 1 Photochemical reaction scheme for the 2-nitrobenzyl cage. *C* caged molecule, *I* activated intermediate, *P* photoreleased product, *SC* spent cage. *Arrowheads* mark the atoms/fragments that will end up in the photoreleased product molecule. The *asterisk* marks the oxygen atom that came from the nitro (NO_2) group in the *ortho* position. *Z* is the moiety that is caged and that will be released upon photolysis. R^1, R^2, and R^3 are substituent groups that can be either H or a more structurally complex moiety. Most commonly, $R^1 = H$, CH_3, or CO_2H, and $R^2/R^3 = H$, OCH_3, or $-O-CH_2-O-$. When R^2/R^3 is one of the non-H substituents, maximal light absorption by the cage shifts to longer wavelengths (from ~260 to ~360 nm). Intermediate *I* has a more conjugated double-bond system (alternating double-single bonds) than *C*; thus the optimal absorption wavelength is longer for *I* than for *C*. *Note*: The following conventions are used in all structural drawings in this chapter: (1) Implicit carbon: Every unlabeled vertex, whether internal or terminal, represents a carbon atom. (2) Implicit hydrogen: Every carbon has a sufficient number of (undrawn) hydrogens to make the total number of bonds to that carbon equal to 4. (3) Explicit heteroatoms: non-carbon, non-hydrogen atoms (e.g., O, N) are labeled explicitly; hydrogens attached to the heteroatom are also explicitly drawn

where *C* symbolizes the intact caged molecule; *I*, the activated intermediate species; and *P*, the photoreleased product. In the process of releasing the desired product, *P*, the photolyzed or "spent" cage (SC) is also generated as a side product. At present, the vast majority of caged molecules that have been made are based on the photochemistry of the 2-nitrobenzyl cage, which is illustrated in Fig. 1.

1.1 The Need for Flash Photolysis

Because the rate at which photorelease occurs is a figure of merit for a caged molecule, the kinetics of the photorelease reaction should be characterized. The most straightforward technique for analyzing the kinetics of a photochemical reaction is flash photolysis. In flash photolysis, a brief flash of actinic light (see Note 1) is delivered to a sample of a caged molecule and the progress of the consequent photochemical reaction is monitored. A useful guideline is that the temporal duration of the actinic flash should be much shorter than the time scale on which photochemical reaction occurs (i.e., the activating effect of the light should be essentially "instantaneous" relative to the consequent molecular changes). This temporal criterion ensures that the observed kinetics of the photochemical reaction are independent of the temporal profile of the actinic flash.

Flash photolysis and kinetic spectroscopy are methods for probing the mechanisms of photochemical reactions. They are essentially different from undertakings that are largely procedural (e.g., a titration, size-exclusion chromatography, or fabrication of liposomes), and their proper application demands chemical knowledge and chemical intuition. Therefore, rather than setting a rigid sequence of steps, it is more useful to provide general guidelines and explanations on the methodology so that it may be applied flexibly.

2 Materials

1. Flash kinetic spectrometer
   Commercial vendors supplying fully configured instruments with digital interface to computers include Edinburgh Instruments, Ltd., and Applied Photophysics, Ltd.

2. High-energy laser mirrors
   Conventional mirrors cannot withstand the high instantaneous power of laser pulses, which can ablate the reflective surface. Mirrors created by deposition of dielectric thin films are the reflectors of choice. Most suppliers have stock mirrors that reflect the most common laser emission wavelengths (e.g., 1,064, 532, 355 nm, etc.). Mirrors are typically specified for either 0° or 45° angle of incidence of the laser beam. The mirrors are readily available from CVI Melles Griot and Newport Corporation.

3. Harmonic separators
   Harmonic separators are dichroic beam splitters based on dielectric thin-film coatings. These are used to separate different wavelength components in a laser beam when multiple harmonics are present in the beam (e.g., for Nd:YAG lasers, commonly available separators can reflect the second harmonic at 532 nm while transmitting the fundamental at 1,064 nm, or reflect the 355-nm third harmonic while transmitting the 532- and 1,064-nm components). CVI Melles Griot and Newport Corporation are commonly used vendors.

4. Beam dump
   Beam dumps or beam stops are available from vendors such as Thorlabs, Newport Corporation, or CVI Melles Griot. Commercial beam dumps can be quite costly, however. An easy, inexpensive, and effective solution is to clamp together a stack of double-edge razor blades (a stack 1–2 cm in thickness is sufficient). Since the thickness of the cutting edge of a razor blade is ~300 nm, essentially no reflection occurs at the edge. The steep spaces between the beveled cutting edges serve as efficient traps for any light that enters (the cutting-edge face of a properly clamped stack of blades should appear completely black—no light is reflected). For reasons of safety, the razor stack should be in an enclosure that allows the laser beam to reach the cutting edges but prevents accidental injury.

5. Optical glass filters

   It is often desirable to block light below a certain wavelength (e.g., UV light in a light beam used for probing a solution sample) while allowing long-wavelength light to pass through. Low-fluorescence, long-pass filters of the KV series are available from Schott North America, Inc. Equivalent filters are available from Oriel Instruments (now a brand of Newport Corporation).

6. Analysis software

   Commercial flash photolysis instruments typically come with basic data analysis routines. Additional software for data analysis include Origin Pro (OriginLab), which has built-in analytical features including nonlinear least-squares curve fitting, and the MATLAB environment (MathWorks), for which various "toolboxes" are available, including one for curve fitting.

7. Caged ATP (NPE-ATP, Molecular Probes).

8. 0.01 M Sodium phosphate buffer, pH 7.4.

3 Methods

3.1 The Basic Flash Photolysis Setup

The basic experimental setup for photolysis is shown schematically in Fig. 2. In a typical experiment, a focused light beam from a lamp (the *probe* beam) is passed through a solution (S) of the caged molecule. A pulse of UV light is delivered into the sample along a path that intersects the probe beam perpendicularly. Transient intermediate species generated by the UV flash absorb light from the probe beam, and the resulting changes in the probe beam

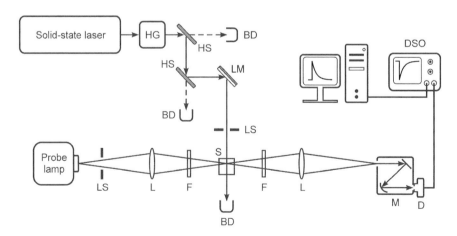

Fig. 2 Schematic representation of flash photolysis-transient absorption spectroscopy. *BD* beam dump, *D* photodetector, *DSO* digital storage oscilloscope, *F* optical filter, *HG* harmonic generator, *HS* harmonic separator, *L* lens, *LM* laser mirror, *LS* laser/lamp shutter, *M* monochromator, *S* sample

intensity are monitored by a photodetector (D). The output of the photodetector is recorded by a digital storage oscilloscope (DSO) or transient waveform recorder, and subsequently stored and processed on a computer. Actinic UV light pulses are generated by passing the infrared (IR) fundamental emission of a solid-state laser through a harmonic generator (HG), the output from which is scrubbed of light at non-UV wavelengths by successive reflection from harmonic separators (HS); the resulting UV pulse can be deflected by a laser mirror (LM) into the sample. Beam dumps (or beam stops) trap undesired components of the laser emission. Shutters (LS) gate both the laser pulse and the probe beam so that the sample is exposed to light only during a measurement. In the probe beam path, appropriate optical filters (F) before the sample block short-wavelength components in the probe beam that can cause undesired photolysis of the sample, and those after the sample prevent any scattered laser flash from reaching the detector. Before reaching the detector, the probe beam is passed through a monochromator (M) so that transient intensity changes can be measured at diverse wavelengths, if desired. Key components of the system will be discussed in greater detail in subsequent section.

3.2 The Actinic Light Source

Historically, flash photolyses were first performed using flash lamps, which delivered light pulses that lasted on the order of a microsecond or longer. Flash-lamp-based systems give excellent data when the photoinduced chemical changes occur on the many-microsecond to millisecond time scale. When photochemical changes occur on a time scale shorter than the flash lamp pulse duration, pulsed laser sources become necessary. Common Q-switched solid-state lasers readily emit pulses with few-nanosecond duration. Mode-locked lasers can give pulses with picosecond duration.

Commonly used solid-state lasers tend to have a gain (lasing) medium consisting of a glass or crystalline material (e.g., silicate or phosphate glass; yttrium aluminum garnet, YAG) that is doped with a rare-earth metal ion (e.g., neodymium, Nd^{3+}; ytterbium, Yb^{3+}). These lasers naturally emit at a fundamental frequency that corresponds to infrared light (e.g., the fundamental emission of Nd:YAG is at 1,064 nm). Currently available caged molecules are photolyzed at UV wavelengths (≤ 400 nm). To use a solid-state laser as the actinic light source, the laser output must be in the UV. This is accomplished by fitting the laser with a harmonic generator comprising nonlinear crystals that can "up-convert" the fundamental emission. Thus, for an Nd:YAG laser, the second, third, and fourth harmonics can be generated at 532 nm, 355 nm, and 266 nm, respectively. Vendors of pulsed solid-state lasers also routinely supply the harmonic generator accessory. For studying most caged molecules, the third harmonic at 355 nm and, less commonly, the fourth harmonic at 266 nm are used. Commercial models provide energies ranging from a few tens to several hundred millijoules per output pulse.

3.3 Isolating the Desired Harmonic Beam for Flash Photolysis

Generating higher harmonics from the fundamental is typically done through *sum frequency generation* in a crystal with nonlinear properties. For an Nd:YAG laser, the fundamental emission is at $\lambda_1 = 1,064$ nm, with equivalent frequency $\upsilon_1 = c/\lambda_1 = (2.9979 \times 10^8 \text{ m/s})/(1,064 \times 10^9 \text{ m}) = 2.8176 \times 10^{14}$ Hz. The beam of photons at υ_1 is first passed through a nonlinear crystal that generates photons at twice the fundamental frequency, or $\upsilon_2 = \upsilon_1 + \upsilon_1 = 5.6352 \times 10^{14}$ Hz, which corresponds to $\lambda_2 = 532$ nm (green light). The emerging beam contains photons at υ_2 as well as unconverted photons at υ_1. This beam is passed through a second nonlinear crystal, in which frequency summing ($\upsilon_3 = \upsilon_1 + \upsilon_2$) again occurs to generate photons at $\upsilon_3 = 8.4528 \times 10^{14}$ Hz, corresponding to $\lambda_3 = 355$ nm (UV light). Since 100% efficiency is never achieved in harmonic generation, the final output beam contains photons at all three wavelengths, λ_1, λ_2, and λ_3. The components at λ_1 and λ_2 can still be at very high intensity, and should be removed before the output beam is used for flash photolysis.

Output at λ_3 can be separated from output at λ_1 and λ_2 by using *harmonic separators* (or *harmonic beam splitters*). These are optical elements that reflects light at one wavelength (or within a narrow wavelength band) while transmitting light at other wavelengths. A pair of harmonic separators working in series is adequate to remove most of the output at λ_1 and λ_2 to leave high-intensity emission at λ_3. A typical arrangement for a pair of harmonic separators is shown in Fig. 3.

3.4 Determining the Temporal Profile of the Actinic Light Flash

An example of a photometric recording of a pulse from a *Q*-switched Nd:YAG laser is shown in Fig. 4a. The trace shows a peak with full width at half-maximum amplitude (FWHM) of 8.5 ns, followed by minor oscillations that die out around 50 ns. Such a trace informs the (essentially arbitrary) decision of how much of the transient absorption data to exclude from analysis because of potential interference from either the actinic light pulse or instabilities in the high-sensitivity photometric circuitry.

An easy way to monitor the temporal profile of the laser pulse is to measure light scattered from the pulse by pure water. A light beam passing through water can undergo elastic (Rayleigh) and inelastic (Raman) scattering by water molecules (Fig. 4b). The Rayleigh scattered photons have the same wavelength as the incident light. The Raman scattering peak is shifted to longer wavelengths (lower energy) because during inelastic scattering, energy equivalent to one vibrational quantum is gained by the scattering water molecule and lost by the scattered photon. Whereas the Rayleigh scattering peak is intense and sharp, the Raman peak is feeble and broad. The amplitudes of both peaks are directly proportional to the incident light intensity and could be used to trace the temporal pulse profile. However, a laser flash is extremely intense and essentially monochromatic (FWHM of a laser emission

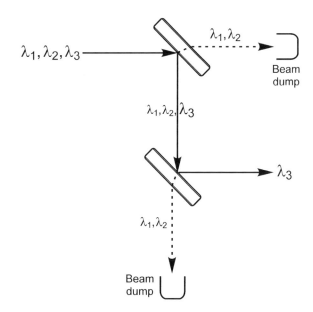

Fig. 3 Using harmonic separators to isolate the third harmonic. The beam containing components at λ_1, λ_2, and λ_3 impinges on the first harmonic separator at an angle of incidence of 45°. Much of the components at λ_1 and λ_2 are transmitted and are discarded in a beam dump. The λ_3 component, along with residual λ_1 and λ_2 components, is reflected, and impinges on the second harmonic separator at 45° incidence. The λ_1 and λ_2 components are transmitted and discarded in a beam dump, while light at λ_3 is reflected, ready for use in flash photolysis. The beam dumps are necessary, because substantial energy remains in the λ_1 and λ_2 beams, which can damage objects and cause burns. The setup, as diagrammed, always reflects, and thus minimizes loss of the λ_3 component (at the expense of spectral purity). Using this setup, the final λ_3 beam still contains minor contamination by light at λ_1 and λ_2, which do not typically interfere with flash photolysis. If higher spectral purity is desired, more harmonic separators can be added to the setup

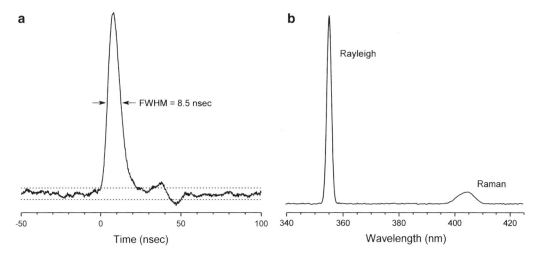

Fig. 4 Temporal profile of third-harmonic (355 nm) pulse from an Nd:YAG laser. (**a**) Photometric recording of the third-harmonic pulse from an Nd:YAG laser (Quanta Ray GCR-18S, Spectra Physics). The full width at half-maximum amplitude (FWHM) is 8.5 ns. There are minor ripples in the trace that subside in tens of nanoseconds. The *dotted lines* delimit the $\pm3\sigma$ limits of the baseline noise. Based on these limits, the trace returns to baseline in ~50 ns. The profile was obtained by measuring the Raman scattering of the laser pulse by pure water (>18 MΩ cm resistivity). (**b**) Rayleigh and Raman scattering from a sample of pure water. A beam of 355-nm light undergoes elastic (Rayleigh) and inelastic (Raman) scattering by water molecules. Light from a xenon arc lamp, passed through a monochromator set at 355 nm, was used for the experiment

peak is <0.1 nm); therefore the Rayleigh peak is extremely intense and narrow. This makes the Rayleigh peak difficult to use (one is either tuned to the peak, with intense scattered light intensity, or off peak, with very little scattering intensity). The Raman peak is much broader, so it is easy to set the detector wavelength within the broad envelope of the Raman peak. In making the scattering measurement, it is important to place a barrier filter in front of the detector that blocks transmission of the Rayleigh scattering but permits transmission of the Raman scattering.

3.5 Monitoring the Progress of the Photochemical Reaction

Reaction Scheme 1 implies that the kinetic progress of the photochemical transformation could be characterized by monitoring the abundance of C, I, or P as a function of time. In principle, any molecular property of C, I, or P that can be measured rapidly could be used to monitor reaction progress. In practice, using a spectrophotometer to measure changes in light absorption is the most common method for tracking the photochemical reaction. According to Beer's Law (see Note 2), absorbance is directly proportional to concentration of the light-absorbing species. Therefore, after an actinic flash, one could use absorbance measurement to monitor the disappearance of the starting caged molecule (C), the appearance and disappearance of the photochemically generated intermediate (I), or the appearance of the photoreleased product (P).

Of the three measurement options, the second (monitoring I spectroscopically) is most commonly used for two reasons. First, as shown in Fig. 1 (and explained in the figure legend), intermediate I can absorb light at longer wavelengths than C; therefore, it is possible to monitor I at a wavelength where C cannot absorb light. Second, under ideal conditions (as implied by Scheme 1), disappearance of I should be concomitant with release of product P, which is the process of greatest practical interest. Of the other possible options, the first, monitoring the consumption of C, is typically not used. This is because measuring the absorbance of C requires C to be exposed to light that it can absorb, and such light would typically cause the same photochemical reaction as the actinic flash. Thus, the very act of observation would steadily destroy C. The third option is to monitor the appearance of P directly. Although this would be ideal, it is rarely possible in practice, because practically all the bioactive molecules that have ever been photoreleased either do not have spectroscopic signatures that can be monitored in the UV-visible wavelength range (e.g., glutamate, GABA, Ca^{2+}, H^+), or have absorption spectra that overlap extensively with the absorption spectrum of the starting caged molecule (e.g., ATP, cAMP, cGMP). Finally, appearance of the spent cage, SC, could be monitored, but it is typical for the UV-visible absorption spectra of SC and C to overlap extensively, making independent monitoring of SC difficult. Figure 5 illustrates the

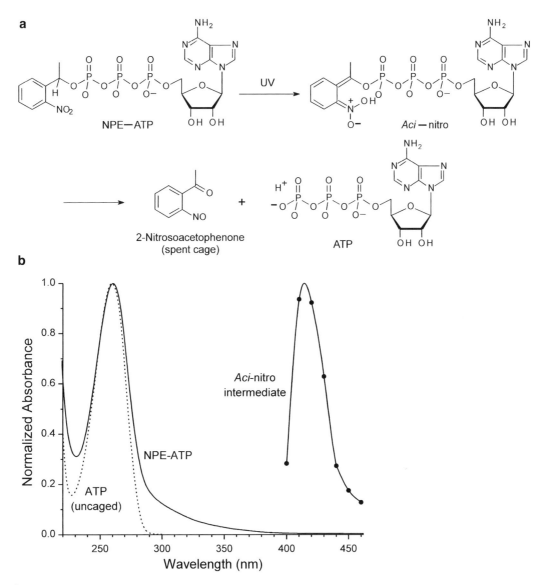

Fig. 5 NPE-caged ATP: Photolysis reaction scheme and UV-visible spectra. (**a**) Photochemical reaction scheme for NPE-caged ATP. (**b**). Normalized UV-visible absorption spectra for NPE-caged ATP, ATP, and the photo-generated *aci*-nitro intermediate. On the *aci*-nitro spectrum, *filled circle* are experimental points obtained by transient absorption spectroscopy, and the *curve* is a cubic spline fit of the data points. Note that whereas the spectra of the caged and uncaged ATP overlap extensively, the spectrum of the intermediate can be measured without interference from the other spectra. All spectra were recorded in 0.01 M sodium phosphate buffer (pH 7.4)

above discussion by showing relevant UV-visible spectra for a caged ATP (NPE-caged ATP).

Sometimes product photorelease is concomitant with the uptake or release of a proton. Therefore, the process can be monitored indirectly by using a colorimetric pH indicator. Not infrequently, the photochemical process changes the total number of ionizable groups in the system (the product species may be more or less ionized than the starting material). Therefore product release may be monitored through electrical conductance changes in the photolyzed solution. These and other means of monitoring the photochemical reaction are discussed in a later section.

3.6 Sample Concentration and Alignment of the Actinic and Probe Light Beams

In a flash photolysis experiment, the actinic flash generates transient intermediates that are monitored spectroscopically by the probe beam. Because the intermediate species are generated at low abundance and are short-lived, the focused probe beam must be carefully aligned to the small volume within which the intermediates are generated by the actinic flash. This optimizes detection of the transient absorbance signal. As a beam of light passes through an absorbing sample, its intensity diminishes exponentially. This is a consequence of Beer's Law (see Note 2). For a solution with an absorbance of A, as a light beam of initial intensity, I_0, travels a distance, x, through the solution, the fraction of light that remains unabsorbed (I/I_0) is given by $I/I_0 = 10^{-Ax} = e^{-2.303Ax}$. This behavior is shown in Fig. 6a for three different values of A. For a solution with $A = 1$, the beam must travel a distance of ~7 mm into the sample before 80% of the light is absorbed. For $A = 2$, that distance is ~3.5 mm, and for $A = 3$, ~2.3 mm. Thus, for the most concentrated sample, with $A = 3$, most of the light is absorbed within the first 2–3 mm—this means that most of the transient intermediate species would be generated within the first 2–3 mm. The probe beam can usually be focused to a spot with diameter of a few mm.

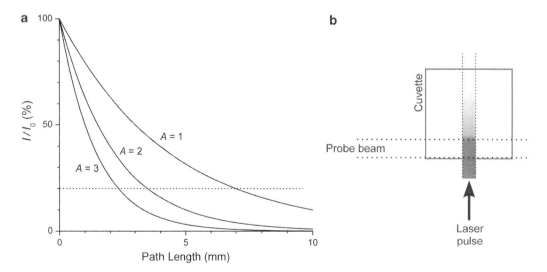

Fig. 6 Sample absorbance and alignment of actinic and probe light beams. (**a**) Attenuation of light beam as function of depth of penetration into an absorbing solution for three different values of absorbance (*A*). *Dotted line marks* 80% attenuation (*I/I₀* = 20%). (**b**) Alignment of the probe beam with the actinic flash relative to the sample cuvette. Attenuation of the laser pulse as it penetrates the sample is represented by sharply diminishing gray levels

To optimize detection of the transient species, the focused probe beam should maximally overlap the region in which most of the actinic light is absorbed (Fig. 6b). As a rough rule of thumb, having a sample with an absorbance between 2 and 3 at the wavelength of the laser flash is adequate. Of course, there are times when this is not possible. Sometimes, a caged compound may be in short supply, or the cost may be prohibitive. In such cases one is forced to use a sample at lower concentration. Reducing caged molecule concentration attenuates the transient absorbance signal but generally does not adversely affect the kinetic measurement.

It should be noted that the laser flash depletes caged molecules in its path. To maximize the amount of caged molecule each successive actinic pulse can convert, the sample solution can be constantly stirred magnetically or, if a flow cell is used, a fresh aliquot of solution could be advanced into the cell to displace the bolus of solution that had been photolyzed. In a stirred sample, as the caged compound in the sample becomes significantly exhausted, the transient absorbance signal caused by the laser flash will diminish noticeably. The exhausted sample can then be replaced with fresh sample. Stirring has a cost: very long-lived photochemical intermediates can be swept out of the volume monitored by the probe beam. This will be discussed in the next section.

3.7 Determining the Duration of Data Acquisition

The photochemical intermediate species is generated upon absorption of light from the laser pulse, and its concentration declines rapidly. The typical transient absorbance change in the flash-photolyzed sample decays exponentially over time. It is advisable during preliminary investigation to monitor the absorbance for a relatively long time (e.g., ~5 ms) after the laser pulse. This immediately reveals the time scale over which the transient species persists. Once this is known, then the actual data acquisition period can be set. As a rule of thumb, for an exponential decay with exponential lifetime, τ, or half-life, $t_{1/2}$ ($t_{1/2} = \ln 2 \cdot \tau = 0.693\tau$), data acquisition should cover a time range $\geq 3\tau$ or $\geq 5 t_{1/2}$ (note that more precisely, $5 t_{1/2} \approx 3.5\tau$; these are rough guidelines). Figure 7 illustrates the rationale. It can be seen that, by definition, the decay is 50% complete after a single $t_{1/2}$ has elapsed, whereas it is ~97% complete after $5 t_{1/2}$. Acquiring data for at least $5 t_{1/2}$ ensures that the decay process is captured almost in its entirety. The more complete data set improves data analysis, which is usually performed through nonlinear least-squares curve fitting to extract τ or $t_{1/2}$. The quality of the curve fitting is sensitive to the baseline or plateau level to which the curve decays; therefore, the more closely the data points at long times approach the baseline, the better the quality of parameters derived from the analysis.

There are conditions under which data acquisition for $5 t_{1/2}$ may not be possible. Since the solution is stirred to restore sample homogeneity before each successive laser pulse, very long-lived

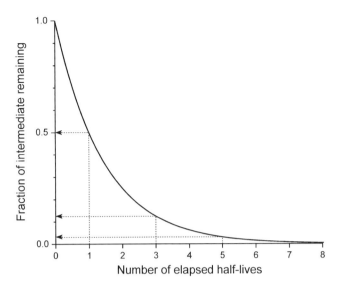

Fig. 7 Completeness of exponential decay as function of number of elapsed half-lives ($t_{1/2}$). *Solid line* shows a single-exponential decay. Three sets of *dotted lines* show the fractions of initially generated transient intermediate that remain after 1, 3, and 5 half-lives have elapsed: After 1 $t_{1/2}$, 50% remains; after 3 $t_{1/2}$, 12.5% remains; after 5 $t_{1/2}$, only 3.1% remains

photochemical intermediates can be swept out of the volume being monitored by the probe light beam. The turbulence caused by stirring is manifested as instability in the absorbance signal at long times. Such artifacts can begin to become noticeable at several milliseconds. This sets the practical upper limit of lifetimes that can be measured in the flash photolysis system.

3.8 Conversion of Light Intensity Data and Signal Averaging

The output of the photodetector is typically a voltage whose magnitude is proportional to the intensity of light impinging on the detector. The raw data is therefore in the form of intensity vs. time. Because only the absorbance of the sample is linearly related to concentration (see Note 2), while the transmitted intensity is not, the raw data must be converted into absorbance vs. time. Examining the definition of absorbance gives the conversion relationship (see Note 3):

$$\Delta A = -\frac{1}{2.303}\frac{\Delta I}{I},$$

where I is the intensity before the laser pulse. This conversion is typically performed automatically on modern instruments.

In many cases, each laser pulse may not generate a large quantity of the transient intermediate species. This means that the transient absorbance change will be small, and the signal-to-noise ratio (SNR or S/N) of the data trace will be poor. Signal averaging is the remedy. Modern instruments typically allow two ways to signal-average. First, light intensity vs. time data recorded after each laser pulse is allowed to sum within the digital storage oscilloscope. When a sufficient number of traces have been summed, a single averaged trace is downloaded into the computer for further processing. Second, each raw intensity data trace collected after a laser pulse is downloaded, converted to absorbance data, and stored on the computer, and the experimenter can average multiple transient absorbance traces after the measurements are complete. Of these two, the second approach is preferable, because automatic summing of raw data in the oscilloscope means that an occasional aberrant raw data trace (containing, for example, severe baseline drift or a discontinuity stemming from a vibration in the instrument) will be averaged with all the good data. The bad data set may introduce no obviously aberrant features into the average trace to announce its presence, but it nonetheless can exert a pernicious effect on subsequent analysis. Figure 8 illustrates this point.

3.9 Nonlinear Least-Squares Curve Fitting

The transient absorbance changes observed after flash photolysis arise from photochemical intermediate species that disappear over time. The goal of data analysis is to extract the kinetic parameter—the exponential time constant τ (or half-life $t_{1/2}$)—that characterizes the time course of disappearance. The usual practice is to fit the data to an exponential decay function, with the form $y = y_0 + A e^{-t/\tau}$, where y_0 is the baseline, and A is the amplitude (Fig. 9a). The parameters A, τ, and y_0 are adjusted through a least-squares procedure to optimize the fit. The objective of curve fitting is to minimize the difference between the data and the fitted curve. More precisely, the quantity that is minimized is "chi-square":

$$\chi^2 = \sum_{i=1}^{N} \frac{\left[y_i - y(t_i) \right]^2}{\sigma_i^2}$$, where $y(t_i)$ is the actual measured value at

time t_i, y_i is the value of the fitted curve at t_i, and σ_i is the uncertainty in $y(t_i)$, which is generally just the standard deviation of the noise in the measurements; the sum is taken over all N time points in the data set. Descriptively, χ^2 is the sum of the *squared* deviations of the actual data points from the fitted curve (weighted by the statistical noise), and finding the minimum of χ^2 leads to the best fit, hence "least squares." When a single-exponential decay component gives a poor fit, the data may be fit to more than one exponential: $y = y_0 + A_1 e^{-t/\tau_1} + A_2 e^{-t/\tau_2} + A_3 e^{-t/\tau_3}$. Generally, reaction mechanisms rarely justify more than three components and, in practice, the noise in the experimental data typically would not allow good fits with more than three components. In order to begin the process of iteratively adjusting the parameters to reduce χ^2, all algorithms require an initial "guess" of the parameter values—rough "eyeball estimates" are usually good enough. When adjustment of the parameters reduces χ^2 by less than a preset tolerance, the fitting process is concluded. Reference 1 is a good foundation for data analysis and curve fitting.

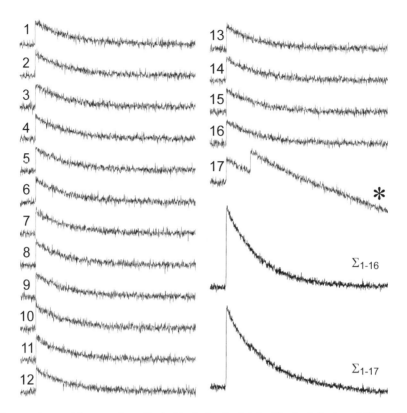

Fig. 8 Signal averaging and the negative effect of including bad data. Stacked display of 16 good traces and a 17th trace that contains a discontinuity and a severe baseline drift (*asterisk*). The two larger traces in the lower left show the results of averaging only the good traces (Σ_{1-16}), and averaging all traces including trace 17 containing artifacts (Σ_{1-17}). Note that signal averaging does improve the signal-to-noise ratio: $S/N = 10$ in the individual traces; $S/N \approx 40$ in the averaged traces (as expected for random noise, S/N improves as $\sqrt{n}$ where n is the total number of traces averaged). Importantly, although trace 17 is wildly aberrant, its inclusion in the average does not produce an average trace that looks grossly wrong. Nonlinear least-squares fits to the averaged traces yielded the following time constants: for Σ_{1-16}, $\tau = 20.06 \pm 0.16$ ($t_{1/2} = 13.90 \pm 0.11$); for Σ_{1-17}, $\tau = 26.59 \pm 0.23$ ($t_{1/2} = 17.74 \pm 0.16$). Both curve fits showed $R = 0.995$, and the residuals showed no obvious systematic deviations from 0. Thus inclusion of 1 aberrant trace in 17 changed the extracted time constants markedly without giving any obvious clues to its presence. These data are simulations: each individual trace comprised normally distributed random noise with standard deviation of $\sigma = 1$ added to a single-exponential decay, $10e^{-t/20}$. Trace 17 is the same as 1–16, but includes an additional signal component: a jump with amplitude 9 at $t = 15$ followed by a linear downward drift with a slope of -0.25882. Each trace consists of 100 points of baseline followed by 1,001 points of transient signal. The starting point of the signal was given a jitter of 0, ± 1, or ± 2 time bins to simulate real experimental recordings

Fig. 9 (continued) decays (*gray dots*) were fit with either a single-exponential component (*thick curve*) or two different exponential components (*thin curve*). The corresponding residual plots are labeled "1-exp" and "2-exp," respectively. When the fit quality is good (2-exp), the residuals are essentially randomly distributed near 0, whereas improper fitting (1-exp) gave residuals that show large systematic deviations from 0. The simulated data comprised (1) the function $y = 10e^{-t/3} + 10e^{-t/20}$, and (2) normally distributed noise with $\sigma = 1$ (thus SNR = 20)

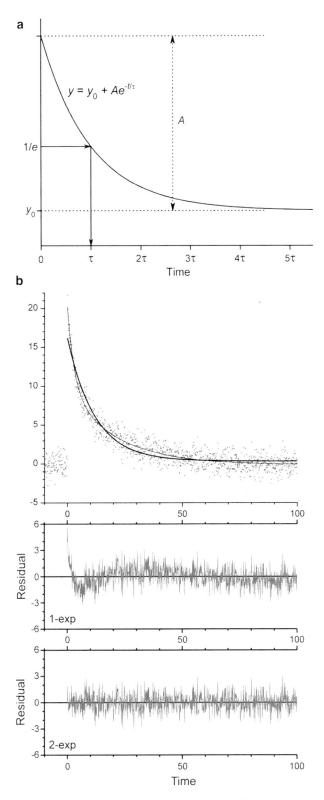

Fig. 9 Goodness of fit of exponential decays. (**a**) Definitions of the parameters characterizing an exponential decay. After a time, τ, has elapsed, the function will have decayed to $1/e$ of its initial value at time 0. (**b**) Data comprising two exponential

Once the best-fit parameters are obtained, there are several ways to check the *goodness of fit*. The simplest and most intuitive is to look at a plot of the residuals—the difference between the fitted curve and the measured data at every time point. If the fitted curve accounts for the temporal behavior of the data perfectly, the residuals should look like random noise. Discernible trends or patterns in the residual plot raise suspicion about the quality of the fit (Fig. 9b). The most widely used quantitative gauge of goodness of fit is the *reduced* χ^2, or $\chi_\nu^2 - \chi^2 / \nu$, where χ^2 is as defined above, and ν, the degrees of freedom, is given by $\nu = N - n$, where N is the total number of data points used for the fit and n is the number of parameters required to define the fitted curve (e.g., the single-exponential function given above has three parameters, y_0, A, and τ, so $n = 3$). Comparing the definitions of χ^2 and χ_ν^2 shows that if the fitted curve describes the data well, then $\chi_\nu^2 \approx 1$; the larger the deviation from 1, the poorer the fit. Another much-used quantitative measure of goodness of fit is the coefficient of determination, $R^2 = 1 - RSS/TSS$. RSS is the residual sum-of-squares, or the residual squared deviation between the fitted curve and the data:

$$RSS = \sum_{i=1}^{N}\left[y_i - y(t_i)\right]^2,$$ where all symbols are as defined in the

previous paragraph. *TSS* is the total sum-of-squares, and represents the total squared deviation between the data and the mean

value of the data ($\bar{y}$, a constant number): $TSS = \sum_{i=1}^{N} N\left[y_i - \bar{y}\right]^2.$

The fraction RSS/TSS is therefore the fraction of the total deviation present in the data that cannot be accounted for by the fitted curve. The smaller RSS/TSS is, the better the fit. In the ideal case, *all* of the deviation from the mean of the data is accounted for by the fitted curve and RSS/TSS = 0, which means $R^2 = 1$. Thus, the closer R^2 approaches 1, the better the fit. It is generally true, however, that as one adds more adjustable parameters to a fit (e.g., increasing the number of exponential components in the fitted curve), unaccounted-for deviation (RSS) will drop, causing R^2 to move closer to 1, but the goodness of fit may not actually improve.

A more useful statistic is the *adjusted R^2*, or $\bar{R}^2 = 1 - \dfrac{RSS(N-1)}{TSS\cdot\nu}.$

The ratio $(N-1)/\nu$ penalizes over-parameterizing the fit. Of course, as the quality of fit improves, $\bar{R}^2$ also increases (see Note 4).

3.10 Alternative Approaches to Monitoring the Progress of the Photochemical Reaction

As mentioned earlier, there are alternative means to monitor the progress of the photochemical reaction. For example, photorelease of product may be concomitant with the release or consumption of a proton, which means that photorelease is accompanied by a change in solution pH. This can be monitored by using a colorimetric pH indicator that shows large changes in its visible-wavelength absorption spectrum in response to pH changes. It is important to choose an indicator whose pK_a is reasonably close to the pH of the experimental solution. Under optimal conditions, an indicator is responsive over a range as wide as $pH \leq pK_a \pm 2$, but careful measurement is required. As a rule of thumb, the indicator should have a pK_a that differs by no more than about ± 1 from the pH of the experimental solution. This ensures that the indicator has sufficient response range left to report actual changes in solution pH. For example, if the sample solution is at pH 7.4, phenol red ($pK_a = 8$) and bromothymol blue ($pK_a = 7.1$) would be suitable; the two indicators show maximal absorbance changes at ~562 nm and ~615 nm, respectively. In order for the reported pH change to be maximal, no significant pH buffering should be added to the sample beyond the small amount afforded by the indicator (and perhaps the caged molecules). The concentration of the indicator used is typically a few tens micromolar to ensure that indicator response is sufficiently fast to report rapid pH changes accurately (see Note 5). Finally, because colorimetric indicators have maximum extinction coefficients on the order of $3–5 \times 10^4$ M^{-1} cm^{-1}, several tens micromolar indicator would give an absorbance of $A > 1$ at the wavelength λ_{max} where the indicator absorbs maximally. Since an $A > 1$ means >90% of the beam is absorbed by the sample, the amount of light reaching the detector may not be sufficient for good signal-to-noise. In such cases, the wavelength of detection can simply be shifted away from λ_{max} so that $A < 1$ at the new wavelength. Some examples of the use of pH indicators can be found in refs. 2 and 3.

In general, photorelease of product may involve a change in the total amount of ionic species in the solution. That is, compared to the starting caged molecule, the photolysis products may carry more or less ionic charge (note that H^+ release or consumption, discussed above, is a special case). Moreover, even if there is no net change in total number of ionic charges, the species present before and after photolysis may have different *ionic mobilities*. In either case, the electrical conductance of the solution should change upon photolysis. Indeed, conductometric recording has long been used to monitor the kinetics of fast reactions (4). It has been used in conjunction with flash photolysis to monitor the progress of photoinduced reactions (5). The basis of fast conductance measurement is a high-frequency AC Wheatstone bridge, one arm of which is the photolyzed sample (6).

Reaction kinetics is inextricably linked to reaction mechanism—the sequence of elementary structural transformations that link reactant to product. Therefore, the ideal technique for monitoring the progress of photochemical reactions is one that can provide detailed information about the structures of the photochemical species. While optical absorption spectroscopy has been most widely used, the measurement reflects the electronic distributions within a

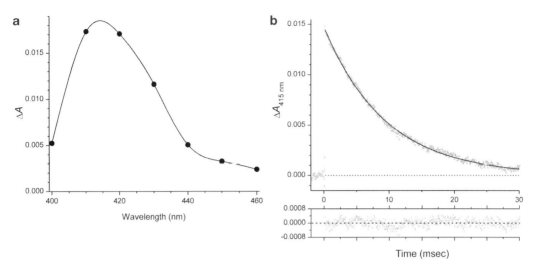

Fig. 10 Flash photolysis of NPE-caged ATP. (**a**) Transient absorption spectrum of NPE-caged ATP. *Filled circles* are data (peak amplitude of transient absorption signal, ΔA, evoked at the indicated wavelengths by a 8.6-ns pulse of 355-nm light). Smooth curve is a cubic spline fit of the data points. The wavelength at which ΔA is maximum is inferred to be ~415 nm. (**b**) Transient absorbance signal acquired at 415 nm (ΔA_{415nm}; *gray points*; average of 30 traces). A single-exponential decay fit to the data gave an exponential time constant of $\tau = 9.44 \pm 0.05$ ms ($t_{1/2} = 6.54 \pm 0.04$ ms). The *lower panel* shows the residuals from the curve fit

molecule but generally does not reveal details of molecular structure. In contrast, infrared (IR) spectroscopy yields information about molecular vibrations, which depend intimately on details of bonding and structure within a molecule. Vibrational spectroscopy is therefore well suited to deciphering photochemical mechanism, although considerable expertise and chemical insight are required. A number of mechanistically informative flash photolysis studies have been conducted using time-resolved IR spectroscopy (7–11).

3.11 Demonstrating the Basic Experiment: Flash Photolysis of Caged ATP

Caged ATP (NPE-caged ATP, or NPE-ATP, or P^3-1-(2-nitrophenyl)ethyl adenosine triphosphate) was invented by Jack Kaplan and was the first caged molecule successfully used in biological research (12). It has been extensively characterized (2, 3, 14). Importantly, it has been demonstrated that the disappearance of the transient *aci*-nitro intermediate is concomitant with ATP release. The basic elements of a flash photolysis study of caged ATP are shown in Fig. 10. NPE-ATP is dissolved in 0.01 M sodium phosphate buffer (pH 7.4) to give an absorbance of 0.2 at 355 nm (this is an instance of cost limiting the concentration of caged molecules used in an experiment). This solution was gently stirred to minimize optical artifacts on long time scales; air was not excluded from the sample. Flash photolysis was effected by a 8.5-ms (FWHM), 200-mJ, 355-nm pulse (third harmonic, Nd:YAG laser, Quanta-Ray GCR-18S, Spectra

Physics). Transient absorption spectroscopy was performed on a commercial spectrometer (LP920, Edinburgh Instruments). Transient absorbance change was monitored at 10-nm intervals from 400 to 460 nm. The resulting transient absorbance spectrum (Fig. 10a) shows a peak at ~415 nm, which was the wavelength at which all subsequent transient absorbance data were acquired. A total of 30 individual intensity traces were acquired in a fresh solution at 25°C, at ~10-s intervals. After conversion to transient absorbance data, the traces were averaged to yield the curve shown in Fig. 10b. The data were well fit by a single-exponential decay with $\tau = 9.44 \pm 0.05$ ms—in good agreement with the findings of Trentham and others (see Note 6).

4 Notes

1. "Actinic" is often used in the photochemistry literature. It means "capable of inducing photochemical reaction."

2. Beer's Law is commonly stated as $A = -\log\dfrac{I}{I_0} = \varepsilon l c$, where A is the *absorbance*, or *optical density* (OD), of the sample, I_0 is the intensity of the light beam impinging on the sample, I is the light intensity after the beam had passed through the light-absorbing sample, ε is the *extinction coefficient* (more formally known as the *molar absorptivity*) of the light-absorbing molecule in the sample, l is the thickness of the sample (the optical *path length*), and c is concentration of the light-absorbing molecule. It is implicit that the measurement is carried out at a specific wavelength, and that ε is a fundamental property of the absorbing molecule at the specified wavelength.

3. Absorbance is defined as $A = -\log\dfrac{I}{I_0}$; converting from base 10 to natural logarithm yields $A = -\dfrac{1}{2.303}\ln\dfrac{I}{I_o}$. Taking the differentials of this expression leads to $dA = -\dfrac{1}{2.303}\dfrac{dI}{I}$, which means $\Delta A = -\dfrac{1}{2.303}\dfrac{\Delta I}{I}$ when ΔA is small. "Small ΔA" means $\Delta A \ll 1$, which is essentially always true in flash photolysis experiments, because the transient photochemical intermediate is never generated in large amounts. Therefore, the transient absorbance change is always quite small.

4. The definitions for χ^2, χ_v^2, R^2, and $\bar{R}^2$ show that they are not actually the square of some other quantity (i.e., there is no such thing as χ, χ_v, R, or $\bar{R}$). They are merely symbolic names for complex expressions.

5. Required indicator concentration can be estimated using the following rationale. The indicator exist in two forms, In^- and HIn, related by the equilibrium

$$In^- + H^+ \underset{k_-}{\overset{k_+}{\rightleftharpoons}} HIn \quad K_a = \frac{k_-}{k_+}$$

where k_+ and k_- are the forward and reverse rate constants, and K_a is the standard acid dissociation equilibrium constant for the indicator. For all known simple proton transfer reactions, the bimolecular rate constant approaches 1×10^{11} M^{-1} s^{-1}. If an indicator has $pK_a = 7$, then $K_a = 10^{-7} \approx k_- / 1 \times 10^{11}$, or $k_- \approx 10^4$ s^{-1}. The indicator's response to a sudden change in $[H^+]$ is governed by a time constant $\tau = (k_+([H^+] + [In^-]) + k_-)^{-1}$ (15). Assuming $pH \approx 7$ ($[H^+] \approx 10^{-7}$ M), then to have $\tau \leq 1$ μs requires $[In^-] \geq 10$ μM. Finally, because colorimetric indicators have maximum extinction coefficients on the order of $3–5 \times 10^4$ M^{-1} cm^{-1}, several tens micromolar indicator would give an absorbance of $A > 1$ at the wavelength λ_{max} where the indicator absorbs maximally. Since an $A > 1$ means >90% of the beam is absorbed by the sample, the amount of light reaching the detector may not be sufficient. In such cases, the wavelength of detection can simply be shifted away from λ_{max} so that $A < 1$ at the new wavelength.

6. Using the calibration curve in Fig. 4 of ref. 2 yields a rate constant for decay of the *aci*-nitro intermediate of $k = 35.5$ s^{-1} at 22°C, pH 7.4 and an ionic strength of $I = 0.2$ M. Reference 13 gave (1) the ionic strength dependence of k as $\log k = 1.67 - 0.445 \log I$, and (2) an activation energy for k of $E_a = 55$ kJ/mol. Our 0.01 M sodium phosphate buffer at pH 7.4 contains 3.86 mM NaH_2PO_4 and 6.13 mM Na_2HPO_4, giving $I = 0.0222$ M. Applying the ionic strength correction gives $k = 94.3$ s^{-1} at our ionic strength at 22°C. Two rate constants, k_1 and k_2, at temperatures T_1 and T_2, respectively, follow the Arrhenius relation: $\ln(k_2/k_1) = (E_a/R)(T_1^{-1} - T_2^{-1})$, where R is the universal gas constant (8.3144 J/mol K). Adjusting for the difference between 22 and 25°C gives the value of $k = 118$ s^{-1}. Finally, $\tau = 1/k = 8.5$ ms, which differs from our determined value by ~10%.

References

1. Bevington PR, Robinson DK (2003) Data reduction and error analysis for the physical sciences. McGraw-Hill, Boston

2. Walker JW, Reid GP, McCray JA, Trentham DR (1988) Photolabile 1-(2-nitrophenyl) phosphate esters of adenine nucleotide analogues. Synthesis and mechanism of photolysis. J Am Chem Soc 110: 7170–7177

3. Zhao J, Gover TD, Muralidharan S, Auston DA, Weinreich D, Kao JPY (2006) Caged vanilloid ligands for activation of TRPV1 receptors by 1- and 2-photon excitation. Biochemistry 45:4915–4926

4. Knoche W (1986) Pressure-jump methods. In: Bernasconi CF (ed) "Investigation of Rates and Mechanisms of Reactions", Part 2. Wiley, New York

5. Bothe E, Dessouki AM, Schulte-Frohlinde D (1980) Rate and mechanism of the ketene hydrolysis in aqueous solution. J Phys Chem 84:3270–3272

6. Bothe E, Janata E (1994) Instrumentation of kinetic spectroscopy—13. a.c.-conductivity measurements at different frequencies in kinetic experiments. Radiat Phys Chem 44:455–458

7. Barth A, Corrie JET, Gradwell MJ, Maeda Y, Mäntele W, Meier T, Trentham DR (1997) Time-resolved infrared spectroscopy of intermediates and products from photolysis of 1-(2-nitrophenyl)ethyl phosphates: reaction of the 2-nitrosoacetophenone byproduct with thiols. J Am Chem Soc 119:4149–4159

8. Cheng Q, Steinmetz MG, Jayaraman V (2002) Photolysis of γ-(α-carboxy-2-nitrobenzyl)-l-glutamic acid investigated in the microsecond time scale by time-resolved FTIR. J Am Chem Soc 124:7676–7677

9. Corrie JE, Barth A, Munasinghe VR, Trentham DR, Hutter MC (2003) Photolytic cleavage of 1-(2-nitrophenyl)ethyl ethers involves two parallel pathways and product release is rate-limited by decomposition of a common hemiacetal intermediate. J Am Chem Soc 125:8546–8554

10. Il'ichev YV, Schwörer MA, Wirz J (2004) Photochemical reaction mechanisms of 2-nitrobenzyl compounds: Methyl ethers and caged ATP. J Am Chem Soc 126:4581–4595

11. Zhu Y, Pavlos CM, Toscano JP, Dore TM (2006) 8-Bromo-7-hydroxyquinoline as a photoremovable protecting group for physiological use: Mechanism and scope. J Am Chem Soc 128:4267–4276

12. Kaplan JH, Forbush B III, Hoffman JF (1978) Rapid photolytic release of adenosine 5'-triphosphate from a protected analogue: utilization by the Na:K pump of human red blood cell ghosts. Biochemistry 17:1929–1935

13. McCray JA, Trentham DR (1989) Properties and uses of photoreactive caged compounds. Annu Rev Biophys Biophys Chem 18: 239–270

14. Corrie JET, Reid GP, Trentham DR, Hursthouse MB, Mazid MH (1992) Synthesis and absolute stereochemistry of the two diastereomers of P^3-1-(2-nitrophenyl)ethyl adenosine triphosphate ('caged ATP'). J Chem Soc Perkin Trans 1:1015–1019

15. Bernasconi CF (1976) Relaxation kinetics. Academic, New York

Chapter 6

Characterization of One- and Two-Photon Photochemical Uncaging Efficiency

Alexandre Specht, Frederic Bolze, Jean Francois Nicoud, and Maurice Goeldner

Abstract

The idea of using light to unleash biologically active compounds from inert precursors (uncaging) was introduced over 30 years ago. Recent efforts prompted the development of photoremovable protecting groups that have increased photochemical efficiencies for one- and two-photon excitation to allow more sophisticated applications. This requires characterization of one- and two-photon photochemical efficiencies of the uncaging processes.

The present chapter focuses on the characterization of one-photon quantum yields and two-photon cross-sections.

Key words Photolabile protecting groups, Uncaging, Photolysis, Quantum yield, Two-photon

1 Introduction

In order to precisely control the dynamic properties of cellular functions and to monitor the complicated biological processes at a desired time or location in intact cells, efficient light-responsive biologically active compounds are needed. Since its introduction a few decades ago (1, 2), the concept of using light-sensitive protecting groups, initially conceived for synthetic organic chemists (3, 4), has found considerable applications in cellular biology and in neuroscience, to produce what are commonly referred to as caged compounds (5–8). This method, which consists of turning "ON," under light irradiation, functionality momentarily masked by chemical attachment of a light-sensitive protecting group (caging group), allowed tremendous breakthroughs for in situ dynamic studies due to the net spatiotemporal control procured.

Recent efforts prompted the development of photoremovable groups that have increased photochemical efficiencies for one-photon near-UV or visible excitation (e.g., the photolytic reaction should

Matthew R. Banghart (ed.), *Chemical Neurobiology: Methods and Protocols*, Methods in Molecular Biology, vol. 995, DOI:10.1007/978-1-62703-345-9_6, © Springer Science+Business Media New York 2013

occur at wavelengths longer than 320 nm to avoid interference with absorbing biological molecules), and for two-photon excitation to allow more sophisticated applications (9–11).

To acquire full utility of the light activation concept, the photolytic reaction should be fast, quantitative (the photolytic process should be the only ongoing photochemical reaction), and efficient for one- or two-photon irradiation depending on biological phenomenon studied. High two-photon action cross-sections are now also important parameters for caged compounds, especially for molecules which need to be photoactivated in live cells and deep in tissue.

The one-photon efficiency of uncaging upon photoirradiation at a given wavelength is directly related to the product of extinction coefficient (in M^{-1} cm^{-1}) and quantum yield of photolysis ($\varepsilon \times \Phi_u$). Caged compounds with high $\varepsilon \times \Phi_u$ need less intense light irradiation for uncaging, which could reduce cell damage caused by strong UV light irradiation. Mostly two different techniques can be used to determine the one-photon uncaging quantum yield:

– Either using an actinometer

– Or more conveniently, by the comparative method using a caged compound with known quantum yield

The two-photon sensitivity of caged compounds can be quantified with the two-photon uncaging action cross-section, usually defined as $\delta_a \times \Phi_u$, which is the product of the two-photon absorption cross-section δ_a and the uncaging quantum yield Φ_u (the same as for one-photon excitation). Two-photon excitation produces excited states that after non-radiative decay are identical to those obtained by classical UV excitation while overcoming major limitations when dealing with biological materials, like spatial resolution, tissue penetration, and phototoxicity. Indeed, a molecule which absorbs classically in the UV region by one-photon excitation will absorb light in the IR region by two-photon excitation, taking advantage of the transparency window of living tissues. Furthermore, the IR light is much less phototoxic than the UV radiations. The two-photon absorption cross-section δ_a is given in Goeppert-Mayer units (1 g = 10^{-50} cm^4 $photon^{-1}$) in honor of Maria Goeppert-Mayer who was the first to predict theoretically the two-photon absorption phenomenon. As the quantum yield is a unit-free physical quantity, the two-photon uncaging action cross-section is also given in GM. We will present here the comparative method, which uses a reference caged compound with known two-photon uncaging action cross-section.

2 Materials

2.1 Light Sources

1. For one-photon quantum yield determination.
   The most widely used sources of UV-visible light for continuous irradiation in laboratory experiments are xenon or

mercury arc lamps, such as a 1,000 W research arc lamp (Muller Elektronik Optik, Germany) or a Hg 1,000 W HBO mercury lamp (Osram, Germany), but one-photon quantum yield determination can only be defined with monochromatic light sources; therefore this type of equipment must be equipped with band-pass filters or a monochromator (Jobin-Yvon, Germany). Light-emitting diodes (LEDs) or organic light-emitting diodes (OLEDs) emitting in the near-UV region are increasingly employed as bright monochromatic light sources due to their low cost and high performance. More powerful laser light sources can also be applied for quantum yield determination, especially for photolytic processes with low efficiencies.

2. For two-photon uncaging action cross-section determination. The determination of two-photon uncaging action cross-section requires necessarily the use of a pulsed infrared laser. The most commonly used lasers for irradiation of caged compounds are mode-locked titanium–sapphire lasers (see Note 1), for which emission wavelength can be tuned from 650 to 1,100 nm. The pulse repetition frequency is in most cases around 70–90 MHz and the pulse duration is typically around 150 fs. An amplified Ti–sapphire laser delivering 50 fs pulses at a 5 kHz repetition rate can also be used. In this case, the illumination times can be reduced to few minutes instead of around 1 h with the laser described before (see Note 2).

The laser intensity at the sample is adjusted with a polarizer and a half wave plate, and measured with a power meter (Spectra Physics). The laser beam is focused at the center of a microcuvette (100 μL quartz microcuvette, Hellma) by a lens (2 cm focal length). The position of the cuvette can be adjusted in order to localize the focal point in the center of the cuvette light pathway (see Note 3).

2.2 Sample Preparation

1. Phosphate buffer 0.1 M at pH 7.4 is prepared with 18 MΩ water from a Milli-Q apparatus. 1 M solution of potassium dihydrogenophosphate (68.04 g in a 500 mL volumetric flask) and a 1 M potassium hydrogenophosphate trihydrate (114.11 g in a 500 mL volumetric flask) are prepared. In a 1 L volumetric flask, 80.2 mL of 1 M K_2HPO_4 and 19.8 mL of KH_2PO_4 are mixed and the flask is adjusted to 1 L (see Note 4).

2. A 10^{-4} M solution of fluorescein is prepared by diluting of 40 mg of fluorescein sodium salt (Aldrich) in 100 mL of 1 M NaOH and adjusted to 1 L in a volumetric flask (see Note 5).

2.3 Standard Preparation

1. One-photon quantum yield characterization.

   (a) By actinometry
   Any liquid-phase actinometer recommended by the IUPAC can be applied (12). In this chapter, *E*-azobenzene (Fluka) will be described for quantum yield determination using 313 nm irradiation.

(b) By comparison

Commercial caged compounds with well-known uncaging quantum yield can be applied. In this chapter the protocol using 1-(2-nitrophenyl)ethyl-ATP (NPE-ATP) ($\Phi = 0.63$) will be described. The reference NPE-ATP was purchased from Jena Bioscience or Tocris Bioscience and used at 10 mM in water (see Note 6).

2. Two-photon uncaging action cross-section determination

The reference CouOAc has been prepared as described in the literature (13). Its spectral characteristics conform to the published ones (see Note 7). The solid (≈ 2 mg) is dissolved in 1 mL of a 1/1 (in vol.) pH 7.4 buffer and acetonitrile mixture.

3. Optical density determination

The ODs are measured on a UVIKON XL spectrophotometer using paired 3.5 mL precision quartz cells (Roth).

2.4 Reaction Progress Measurement by HPLC

Aliquots of samples (50 μL) were injected into Waters 600E HPLC carried out on an Acclaim C18 column (4.6 × 300 mm); elution was performed at a flow rate of 1 mL/min with a linear gradient of acetonitrile in an aqueous solution of TFA (0.1%) from 0 to 100% (v/v) over 30 min. The compounds were detected by a Waters 2,996 PDA detector operating between 200 and 600 nm.

3   Methods

3.1 One-Photon Quantum Yield Characterization

3.1.1 By Actinometry

1. The photon flow Ep for azobenzene irradiation at 313 nm is defined by

$$Ep = 5.30 \times 10^{-6} \Delta A_{358} / \Delta t \left[\text{Einstein cm}^{-2} \text{s}^{-1} \right]$$

where ΔA_{358} is the change in absorbance at 358 nm and Δt is the irradiation time in seconds.

2. The graph change in the concentration of the studied compound versus photolysis time (in seconds) is plotted. A linear regression is performed in the linear range (photolysis <20%) to determine the reaction rate $\Delta c/\Delta t$ (in mol L^{-1} s^{-1}).

3. The uncaging quantum yield is obtained by using the equation

$$\Phi_u = \Delta c / \Delta t \, Ep^{-1} V$$

where V is the volume of the irradiated solution (3.5 mL).

The quantum yield Φ_u, defined as the ratio of caged molecules converted to the amount of absorbed photons, follows the rate constants of uncaging induced by a given monochromatic light source. Therefore, the reaction progress is measured by analytical methods such as HPLC. The photodecomposition of the starting material can easily be followed by HPLC and can be correlated to the uncaging quantum yield if a stoichiometric release of the biomolecule has been previously observed. The amount of photons absorbed is in practice measured by actinometry. An actinometer is a chemical system that undergoes a light-induced reaction, for which the quantum yield is accurately known.

Preparation of Samples

The reference azobenzene (at 0.64 mM) and the caged compound under study are separately dissolved in methanol and in phosphate buffer, respectively, and then adjusted to equal optical densities at 313 nm. The light source is a 1,000 W research arc lamp with a Hg 1,000 W HBO mercury lamp equipped with a monochromator. The light beam is focused at the center of a cuvette by a lens. The ODs were measured on a UVIKON XL spectrophotometer using paired 3.5 mL precision quartz cells.

Irradiations

1. The quartz cuvette is first filled with 3.5 mL of a 0.64 mM solution E-azobenzene in methanol and stored in the dark.

2. The first step is to determine the photon flow of the irradiation source. Typically, the change in absorbance at 358 nm of E-azobenzene is monitored in function of the irradiation period (see Note 8).

3. The second step is to irradiate 3.5 mL of the caged compound under study at identical optical densities at the specific wavelength of irradiation (313 nm), in phosphate 0.1 M phosphate buffer, and to determine the reaction rate $\Delta c/\Delta t$ (in mol L^{-1} s^{-1}) where Δc is the change in concentration of the caged compound under study and Δt is the irradiation time in seconds. Therefore, 100 µL samples are collected after define irradiation time and immediately frozen in liquid nitrogen for HPLC analysis.

4. Each irradiated sample is analyzed by HPLC to determine the percentage of unreacted product.

Results

1. The photon flow Ep for azobenzene irradiation at 313 nm is defined by

$$Ep = 5.30 \times 10^{-6} \Delta A_{358} / \Delta t \left[\text{Einstein cm}^{-2} \text{s}^{-1} \right]$$

where ΔA_{358} is the change in absorbance at 358 nm and Δt is the irradiation time in seconds.

2. The graph change in the concentration of the studied compound versus photolysis time (in second) is plotted. A linear regression is performed in the linear range (photolysis <20%) to determine the reaction rate $\Delta c/\Delta t$ (in mol L^{-1} s^{-1}).

3. The uncaging quantum yield is obtained by using the equation

$$\Phi_u = \Delta c / \Delta t \, Ep^{-1} V$$

where V is the volume of the irradiated solution (3.5 mL).

By Comparison with Known Caged Compounds

The quantum yield of uncaging can be determined more easily by comparison with a caged compound having a well-defined quantum yield. Therefore the known and unknown caged compounds were co-irradiated at defined concentrations leading to identical optical densities at the specific wavelength of irradiation (see Note 8).

Preparation of Samples

The reference NPE-ATP (at 0.2 mM) and the caged compound under study are separately dissolved in an phosphate buffer and adjusted to equal optical densities at the used one-photon excitation. The light source used is a 1,000 W research arc lamp with a Hg 1,000 W HBO mercury lamp equipped with a monochromator. The light beam is focused at the center of a cuvette by a lens. The ODs were measured on a UVIKON XL spectrophotometer using paired 3.5 mL precision quartz cells.

Irradiations

1. The quartz cuvette is first filled with 3.5 mL of a 0.2 mM solution of NPE-ATP in a pH = 7.2, 0.1 M phosphate buffer.

2. The first step is to verify the uncaging rate of caged NPE-ATP versus light exposure duration. Typically, the uncaging is quantified by HPLC after 1–20 min of irradiation at 315 nm using a 1,000 W mercury arc lamp equipped with a monochromator.

3. The time of irradiation is then adjusted in order to analyze a linear regression (% of photolysis <20%) for NPE-ATP.

4. The second step is to irradiate a mixture of 1.75 mL of a 0.2 mM NPE-ATP and 1.75 mL of the caged compound under study at identical optical densities at the specific wavelength of irradiation (315 nm), in phosphate 0.1 M phosphate buffer. After defined irradiation times (as determined in the first part), more than five 100 μL samples are collected and immediately frozen in liquid nitrogen for HPLC analysis.

5. Each irradiated sample is analyzed by HPLC to determine the percentage of unreacted product.

Results

The graph of % of the remaining compound versus photolysis time is plotted for the studied substance and for the NPE-ATP reference. A linear regression is performed in the linear range (typically % of photolysis <20%), to limit as much as possible errors due to undesired light absorption during photolysis and to access Grad(Substance) and Grad(Reference).

The quantum yield is given by the formula (see Note 9):

$$\Phi_u(X) = \Phi_u(\text{Reference}) \times \text{Grad}[\text{Substance}]/\text{Grad}[\text{Reference}]$$

Where $\Phi_u(\text{Reference}) = 0.63$ for NPE-ATP

3.2 Two-Photon Uncaging Action Cross-Section Determination

The method described here allows the determination of the two-photon uncaging action cross-section $\delta_a \times \Phi_u$ (where δ_a is the two-photon absorption cross-section and Φ_u the uncaging quantum yield). It is based on the irradiation under identical conditions of the compound of interest and a known reference. This method enables to use any caged compound being fluorescent or not.

3.2.1 Preparation of Samples

The reference (CouOAc) and the caged compound to study are diluted in an acetonitrile-phosphate buffer (1/1 in volume) and adjusted to equal optical densities (OD) at half of the two-photon excitation (see Note 10). The ODs in the order of 0.400 were measured on a CARY 4 spectrophotometer using paired precision quartz cells.

3.2.2 Irradiations

1. The microcuvette is first filled with 100 μL of a 10^{-4} M solution of fluorescein in 0.1 M NaOH.

2. The focal point is adjusted to the center of the microcuvette with the aid of fluorescein two-photon excited fluorescence by translating the cuvette support. The microcuvette is then washed thoroughly with water and acetone and allowed to dry.

3. 80 μL of phosphate buffer-acetonitrile solution (1:1 in volume) of CouOAc is placed in the microcuvette.

4. The first step is to verify the quadratic dependence of uncaging rate versus laser power both for the reference compound. Typically, the uncaging is quantified by HPLC after 40 min of irradiation with laser powers P of 50, 100, 150, 200, and 250 mW. The graph of the % of photolysis versus P^2 should be linear. These values have to be adapted according to the two-photon efficiency of the studied substance.

5. The second step is to verify the quadratic dependence of uncaging rate versus laser power for the studied substance in the same way as for four.

6. The irradiation power for the two-photon uncaging action cross-section determination is chosen in the quadratic range to obtain 20% of photolysis after 30–40 min.

7. The third step is to irradiate 80 μL of a phosphate buffer-acetonitrile solution (1:1 in volume) of the studied compound placed in the microcuvette at chosen laser power (step 6) during various durations (for example 0, 10, 20, 30, 40, 50, 60 min). After irradiation, samples are immediately frozen in liquid nitrogen for HPLC analysis.

8. The same experiment is then performed with the reference CouOAc.

9. Each irradiated sample (the CouOAc reference and the studied compound) is analyzed by HPLC to determine the percentage of unreacted product.

3.2.3 Results

1. The graph of the percentage of the remaining compound versus photolysis time is plotted for the studied substance and for the CouOAc reference (Fig. 1). A linear regression is

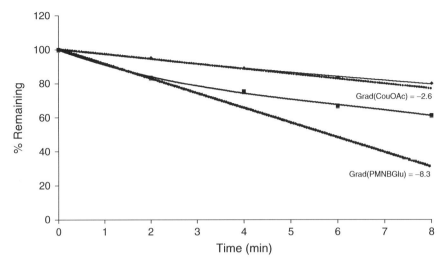

Fig. 1 Example of determination of the two-photon uncaging action cross-section of PMNB-Glu at 800 nm using CouOAc as a reference (irradiations by an amplified laser). In this case, $\delta_a\Phi_u$(PMNB-Glu) = $\delta_a\Phi_u$ (CouOAc) × Grad[PMNB-Glu]/Grad[CouOAc] where $\delta_a\Phi_u$(CouOAc) = 0.13 GM, so $\delta_a\Phi_u$(PMNB-Glu) = 0.13 × (−8.3)/ (−2.6) = 0.41 GM

performed in the linear range (typically % of photolysis <20%) to access Grad(Substance) and Grad(Reference).

2. The two-photon uncaging action cross-section of the studied caged compound is given by the formula (see Note 9):

$$\delta_a\Phi_u(\text{Substance}) = \delta_a\Phi_u(\text{Reference}) \times \text{Grad}\left[\text{Substance}\right] / \text{Grad}\left[\text{Reference}\right].$$

4 Notes

1. Other lasers can be used, such as ns pulsed laser. In this case the pulse duration is longer, providing a smaller instantaneous power, which can lead to longer irradiation times. In all cases, the photolysis has to be tested with a reference compound to determine the time needed for around 20% of photolysis.

2. Picture of the experimental setup for the two-photon uncaging action cross-sections.

3. This solution is stored at 4°C, and has to be discarded after 1 month. The buffer solution is used at room temperature.

4. This solution is stable for a month and stored at 4°C.

5. The solution is stable for at least 12 months at −20°C.

6. The solid form is stable for at least 12 months at −20°C.

7. Each irradiation time should cause a change in A_{358} of about 0.02 and the final value should be kept between 0.85 and 0.9.

8. This easier protocol can show some pitfalls leading to less accurate values especially if the two caged compounds show very different photochemical efficiencies at the wavelength of irradiation.

9. At least three independent measurements concording at 0.2 have to be performed, and the average has to be done to give a two-photon uncaging action cross-section with an accuracy of around 20%.

10. For example, for a measurement at 740 nm of the two-photon uncaging action cross-section, the OD is adjusted to 0.400 at 370 nm.

Acknowledgments

This work was supported by the Université de Strasbourg, the CNRS, and ANR (Grant No. PCV 07 1–0035) and HFSP Young Investigator's Award (RGY0069/2006).

References

1. Engels J, Schlaeger EJ (1977) Synthesis, structure, and reactivity of adenosine cyclic 3′,5′-phosphate benzyl triesters. J Med Chem 20(7):907–911

2. Kaplan JH, Forbush B, Hoffman JF (1978) Rapid photolytic release of adenosine 5′-triphosphate from a protected analogue: utilization by the Na:K pump of human red blood cell ghosts. Biochemistry 17(10):1929–1935

3. Bochet C (2002) Photolabile protecting groups and linkers. J Chem Soc Perkin Trans 1(2):125–152

4. Klan P, Solomek T, Bochet C, Blanc A, Givens R, Rubina M, Popik V, Kostikov A, Wirz J (2013) Photoremovable protecting groups in chemistry and biology: reaction mechanisms and efficacy. Chem Rev 115(1):119–191

5. Marriott G (ed) (1998) Caged compounds. Methods in enzymology, vol 291. Academic, New York

6. Goeldner M, Givens R (eds) (2005) Dynamic studies in biology, phototriggers, photoswitches and caged biomolecules. Wiley-VCH, Weinheim

7. Mayer G, Heckel A (2006) Biologically active molecules with a 'light switch'. Angew Chem Int Ed Engl 45(30):4900–4921

8. Ellis-Davies GC (2007) Caged compounds: photorelease technology for control of cellular chemistry and physiology. Nat Methods 4(8):619–628

9. Lee HM, Larson DR, Lawrence DS (2009) Illuminating the chemistry of life: design, synthesis, and applications of "caged" and related photoresponsive compounds. ACS Chem Biol 4(6):409–427

10. Specht A, Bolze F, Omran Z, Nicoud JF, Goeldner M (2009) Photochemical tools to study dynamic biological processes. HFSP J 3(4):255–264

11. Brieke C, Rohrbach F, Gottschalk A, Mayer G, Heckel A (2012) Light-controlled tools. Angew Chem Int Ed Engl 51(34):8446–8476

12. Kuhn HJ, Bralavsky SE, Schmidt R (1989) Chemical actinometry. Pure Appl Chem 61(2):187–210

13. Furuta T, Wang SS, Dantzker JL, Dore TM, Bybee WJ, Callaway EM, Denk W, Tsien RY (1999) Brominated 7-hydroxycoumarin-4-ylmethyls: photolabile protecting groups with biologically useful cross-sections for two photon photolysis. Proc Natl Acad Sci U S A 96:1193–1200

Chapter 7

Photochromic Potassium Channel Blockers: Design and Electrophysiological Characterization

Alexandre Mourot, Timm Fehrentz, and Richard H. Kramer

Abstract

Voltage-gated potassium (K_v) channels are membrane proteins that open a selective pore upon membrane depolarization, allowing K^+ ions to flow down their electrochemical gradient. In neurons, K_v channels play a key role in repolarizing the membrane potential during the falling phase of the action potential, often resulting in an after hyperpolarization. Opening of K_v channels results in a decrease of cellular excitability, whereas closing (or pharmacological block) has the opposite effect, increased excitability. We have developed a series of photosensitive blockers for K_v channels that enable reversible, optical regulation of potassium ion flow. Such molecules can be used for remote control of neuronal excitability using light as an on/off switch. Here we describe the design and electrophysiological characterization of photochromic blockers of ion channels. Our focus is on K_v channels but in principle, the techniques described here can be applied to other ion channels and signaling proteins.

Key words Azobenzene, Photoswitch, Photochromic ligand, Ion channel, Photopharmacology, Quaternary ammonium compounds, Electrophysiology

1 Introduction

Photochromic ligands (PCLs) are small, freely diffusing molecules that can be reversibly interconverted between two isomeric forms using different wavelengths of light (Fig. 1a). The two isomers bind their protein target with different affinities, making it possible to control binding events (and resulting signaling cascades) reversibly with light. The ligand can be an agonist, an antagonist, or an active site inhibitor, allowing one to turn the protein function on and off repeatedly with different wavelengths of light. Light as a stimulus has numerous advantages, including spatial and temporal precision, non-invasiveness, and orthogonality (most cells are not naturally light-responsive) (1, 2).

In an attempt to manipulate neuronal activity with light, we decided to engineer PCLs for K^+ channels, a class of ion channels that are crucial for setting the resting membrane potential and

Matthew R. Banghart (ed.), *Chemical Neurobiology: Methods and Protocols*, Methods in Molecular Biology, vol. 995,
DOI 10.1007/978-1-62703-345-9_7, © Springer Science+Business Media New York 2013

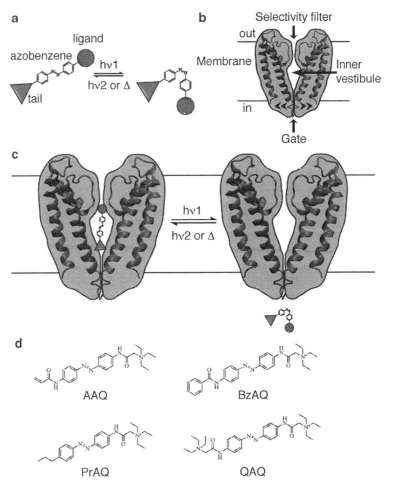

Fig. 1 Photochromic ligands (PCLs) for K⁺ channels. (**a**) Schematic representation of a PCL containing a central photoisomerizable azobenzene core, a ligand head, and a tail. Isomerization between the elongated *trans* and the bent *cis* azobenzenes occurs upon irradiation with different wavelengths of light (hv1 and hv2). The *cis* isomer is of higher energy and can relax back to *trans* spontaneously in the dark (Δ). (**b**) Architectural features of K⁺ channels. K⁺ channels are tetrameric proteins but only two subunits are shown for clarity. The ion path has distinct elements: an external filter that is highly selective for K⁺ ions, a central vestibule, and an intracellular gate. Voltage-gated K⁺ channels have an additional voltage-sensing domain that opens the gate upon membrane depolarization (not depicted here). (**c**) Schematic view of a photochromic blocker for K⁺ channels that blocks the channel in the *trans* form (inner vestibule) and unbinds in the *cis* form. (**d**) Examples of PCLs for K⁺ channels: *AAQ* acrylamid azobenzene quaternary ammonium, *BzAQ* benzylamide azobenzene quaternary ammonium, *PrAQ* propyl azobenzene quaternary ammonium, and *QAQ* quaternary ammonium azobenzene quaternary ammonium

shaping action potentials in neurons and other excitable cells. K$^+$ channels are made of four subunits arranged around a central ion-conducting pore, which contains a selectivity filter, an inner vestibule, and an intracellular gate (Fig. 1b) (3). The inner vestibule is particularly important from a pharmacological point of view, as it constitutes the binding pocket for a class of molecules called open-channel blockers. Tetraethylammonium (TEA) and other quaternary ammonium (QA)-containing molecules prevent ion flow through the channel by sneaking into the inner vestibule after opening of the channel gate (4). Because of their permanent positive charge, most QAs are membrane impermeant and have to be brought inside the cell by artificial means (i.e., the patch pipette). In contrast, some hydrophobic QAs can penetrate cells by passively crossing the lipid bilayer and tend to be trapped inside cells for long periods of time (5, 6).

We have designed a small library of PCLs for K$^+$ channels that have a fixed ligand head group (a QA), an azobenzene photoswitch linker, and a variable tail group (7–10). These PCLs are designed to bind to the inner vestibule of K$^+$ channels in only one of the two configurations (Fig. 1c), light being used to toggle the ligand in and out of its binding pocket. The tail's chemical variability (see Fig. 1d for a few examples) allows for fine-tuning of the PCL's physicochemical, pharmacological, as well as photochemical properties. BzAQ for example is ten times more potent than AAQ, presumably due to better membrane permeation. PrAQ is a *cis* blocker (blocks predominantly in the *cis* configuration), which is an advantageous feature because the compound does not block the channel in the dark. QAQ does not cross the membrane and can be injected into cells through a micropipette to photosensitize a single cell and afford subcellular control of action potential propagation.

In this chapter we describe the design, handling, and use of PCLs to control current through K_v channels. We are focusing on photochromic blockers for K_v channels, but the general strategy is applicable to different types of proteins, as already shown for ion channels (7–13), metabotropic receptors (14), and enzymes (15–18).

2 Materials

2.1 Cell Line

1. Human embryonic kidney (HEK293) cells are a popular choice among electrophysiologists due to multiple attributes: They are easy to grow and maintain, can be transfected with high efficiency, and produce high quantities of proteins. In addition, background K_v channel current from endogenous channels is small (19).

2. The *Xenopus laevis* oocyte is another classical cell of choice for electrophysiologists (20) but its opacity limits optical control and is therefore not recommended for studying light-regulation of ionic currents.

2.2 K_v Channel Clones

The Shaker K^+ channel from *Drosophila melanogaster* was the first K^+ channel to be cloned (21) and has since become an archetypical channel for biophysical and biochemical studies. It belongs to the K_v1 subfamily and shows rapid inactivation, due to the presence of a peptidic "ball and chain" N-terminus. The "ball and chain" (or N-type) inactivation can be removed by deleting amino acid residues 6–46 (Shaker-IR or inactivation removed) (22). Screening of PCLs for K_v channels was performed on Shaker-IR, since the presence of a "ball and chain" may interfere with PCL block (7, 8).

2.3 Chemicals

All chemicals were purchased from Sigma-Aldrich (St. Louis, MO, USA) unless specified otherwise.

All photochromic blockers were synthesized as previously described (7).

2.4 Cell Culture

1. HEK293 medium: Dulbecco's Modified Eagle Medium containing high glucose, glutamine, and sodium pyruvate (D-MEM, Gibco) supplemented with 10% Fetal Bovine Serum (FBS Gibco 26140). Store at 4 °C.

2. Phosphate buffered saline (PBS), pH 7.4 (Gibco). Store at RT.

3. Trypsin 0.25% with EDTA (Gibco); 1 ml aliquots, store at –20 °C.

4. Poly L-lysine (10 mg/ml); 1 ml aliquots, store at –20 °C.

5. Glass coverslips, 12 mm diameter (Fisher).

6. Borate buffer: 50.1 mM boric acid, 26.5 mM sodium tetraborate. pH does not need to be adjusted. Filter-sterilize (0.2 μm) and store at 4 °C.

7. 24-well disposable tissue culture plates with flat bottom (Nunc multidishes, Nunclon).

8. Nitric acid.

9. 12.1 N Hydrochloric acid.

2.5 Transfection

1. HEPES-buffered saline solution (2×-HeBS): 274 mM NaCl, 10 mM KCl, 1.4 mM $Na_2HPO_4 \cdot 7H_2O$, 15 mM glucose, 42 mM HEPES free acid. Adjust pH to exactly 7.12 with NaOH. Filter-sterilize and store at 4 °C. Do not freeze. Make fresh every month.

2. 2.5 M $CaCl_2$. Filter-sterilize and store at 4 °C.

2.6 Recording Solutions

1. For whole-cell recordings.

External solution: 138 mM NaCl, 1.5 mM KCl, 1.2 mM $MgCl_2$, 2.5 mM $CaCl_2$, 5 mM HEPES free acid, and 10 mM glucose. pH is adjusted to 7.4 with NaOH. Store at 4 °C.

Internal solution: 10 mM NaCl, 135 mM K⁺ gluconate, 10 mM HEPES free acid, 2 mM MgCl₂, 2 mM MgATP, 1 mM EGTA. pH is adjusted to 7.4 with KOH. Filter-sterilize solution (0.2 μm) and store 1 ml aliquots at –20 °C.

2. For inside-out patches.

Bath solution: 160 mM KCl, 0.5 mM MgCl₂, 1 mM EGTA, 20 mM HEPES free acid. pH is adjusted to 7.4 with KOH. Store at 4 °C

Pipette solution: 150 mM NaCl, 10 mM KCl, 10 mM HEPES free acid, 1 mM MgCl₂, 3 mM CaCl₂. pH is adjusted to 7.4 with NaOH. Filter-sterilize solution (0.2 μm) and store 1 ml aliquots at –20 °C.

2.7 Equipment

The list of equipment here below is purely informative; other models are compatible too.

1. Spectrophotometer: SmartSpec Plus (Bio-Rad) or NanoDrop 1000 (Thermo scientific).

2. Pipette puller (Sutter P97).

3. Polisher-microforge (Narishige MF830).

4. Capillaries: Borosilicate thin wall with filament (Warner G150TF-3).

5. Syringe needle for filling micropipettes (Microfil 28 gauge MF28G, World Precision Instruments, Inc.).

6. Faraday cage located in a dark room or covered with a black curtain (see Note 1).

7. Air table.

8. Amplifier Axopatch 200A (Molecular Devices).

9. Digitizer: Digidata 1,200 interface (Molecular Devices).

10. Recording/analysis software: pClamp software (Molecular Devices).

11. Microscope: Preferentially inverted.

12. Objectives: 20 or 40×, with high UV light transmission (Nikon Plan Fluor series).

13. Filter cube for GFP fluorescence (GFP-3035B Semrock).

14. Total reflector mirror (Omega opticals XF125).

15. Light source: There are four main possibilities: (1) monochromator (polychrome V, Till photonics, see Note 2); (2) Lambda DG4 (Sutter Instruments); (3) Xenon lamp (175 W or higher) in combination with a filter wheel controlled by a Lambda 10–2 (Sutter Instruments) and narrow band-pass filters (380BP10 and 500BP5); and (4) high-power microscope-LED with corresponding dichroic beam splitters (Prizmatix).

All light sources can be controlled by pClamp and provide light intensities compatible with PCL photoswitching (see Note 3). LEDs are less flexible in terms of action spectrum (need to buy a new LED for any new wavelength needed) but are definitely a cheap alternative.

16. Optic fiber (UV/Vis quartz fiber) connecting the light source to the microscope.

17. Power meter (Newport 840-C).

18. Recording chamber for 12 mm coverslips (Warner RC25).

19. Recording chamber for inside-out patches (Warner RC28).

20. Micromanipulator (Sutter MP285).

21. Perfusion system (Warner VC-8).

3 Methods

3.1 PCL Design

The general design of a PCL can be considered as the structural sum of two or three components: a head (ligand), a photoisomerizable core (photoswitch), and an optional tail (Fig. 1a).

1. Ligand: The ligand can be an agonist (11, 13), a competitive antagonist (14, 23), a pore blocker (7–10, 12), a substrate (16), an inhibitor (15, 17, 18), or even an allosteric modulator, depending on the protein target and the experimental use.

2. Photoswitch: Among the photosensitive groups available, azobenzene has been predominantly used (7–18, 23) due to its advantageous chemical and photochemical properties. Indeed, azobenzene is a small, relatively simple molecule, easy to synthesize and derivatize. The two isomers have significantly different absorption spectra, making it possible to accumulate up to 95% of one isomer under appropriate light conditions (see Note 3). Azobenzenes can be switched back and forth quickly numerous times (no photobleaching) with particular wavelengths of light, typically 360–380 for converting to *cis* (see Note 4) and 460–500 nm for switching back to *trans*. The *cis* isomer also converts back to the low-energy *trans* form slowly in the dark, which implies that in darkness azobenzenes exist almost exclusively in the *trans* form. Importantly, *trans* to *cis* photoisomerization can affect the docking of the PCL into the receptor-binding pocket in two ways: through a dramatic change in geometry (planar to bent) and a slight increase in polarity (~3 Debye).

3. Position for photoswitch attachment: For the ligand, check published structure–activity relationship or crystallographic structures when available to identify positions where a chemical extension is well tolerated by the receptor. Concerning the

photoswitch, the *para* position on the azobenzene is probably best suited to maximize geometrical change upon photoisomerization, although *ortho-* and *meta*-substituted analogues have also been successfully used (11, 23).

4. Linker: The linker between the photoswitch and the ligand should be as short and as rigid as possible, to maximize the deformation endured by the molecule during photoisomerization, and therefore the dynamic range of affinities between the two isomers. Methylene and amide bonds are good starting choices. Care should be taken with substituants in *para* position of the azobenzene as their chemical nature can drastically alter the photochemical properties of the photoswitch (this is also true for *meta* and *ortho positions*). Indeed, azobenzenes that are *para*-substituted with an electron-donating group (i.e., amino) usually relax back to *trans* much more rapidly, which may impair accumulation of the *cis* analogue, and are rather used as fluorophore or quencher (e.g., DABCYL (24)).

5. Tail: In theory, addition of a tail will maximize geometrical change between *trans* and *cis* forms, which may result in a greater shift in affinity upon light isomerization. In practice however, experiment will be necessary to reveal if a tail is required as for K+ channels (7) or not tolerated as for glutamate receptors (13). Interestingly, the tail structure may also dictate whether the molecule is active in *cis* or in *trans* (7, 10) (see Note 5). One should therefore consider making a series of PCLs with (and without) various tails, as it may be hard to predict PCL pharmacological profiles on a new system.

3.2 Cell Culture

1. Coverslip acid wash: Incubate coverslips in nitric acid overnight. Remove nitric acid and incubate in hydrochloric acid another night. Wash extensively with water. Wash with 100% ethanol. Store coverslips in 100% ethanol.

2. Coverslip coating: The day before transfection, flame coverslips and place one coverslip in each well of the culture dish. Add 500 µl of poly-L-lysine (100 µg µl^{-1} dissolved in borate buffer) in each well.

3. Cell plating: Wash poly-L-lysine-treated coverslips four times with 500 µl of water. Trypsinize and plate HEK293 cells at a density of 20–30 thousands cells per coverslip (500 µl of medium per well). Give them enough time to settle back down before transfecting (5–12 h).

4. Transfection: Cells are transfected with K_v channel cDNA (0.5–2 µg) using the calcium–phosphate precipitation method, which is a very efficient and low-cost means to introduce DNA into cells (25). Co-transfection with a fluorescent reporter (eGFP) is usually required (0.05–0.1 µg of DNA) except if the

channel is fused to a fluorescent reporter, or if it is contained with GFP in a mammalian bicistronic expression vector (e.g., pIRES).

Prepare the precipitates: For each well put 11 μl of 2×-HeBS in a sterile 1.5 ml tube. In another tube mix the DNAs (0.5–2 μg for K_v + 0.05–0.1 μg for eGFP) with $CaCl_2$ 2.5 M (1.1 μl) and add water to a final volume of 11 μl. Add slowly the DNA-$CaCl_2$ mix to the HeBS solution and mix gently (no vortex). Let sit for 20 min. Add the transfection mix to the well (22 μl per well) in dropwise fashion, and shake the plate gently. Place in the incubator overnight (37 °C, 7% CO_2). Replace media the next day with 500 μl of fresh HEK293 medium. Electrophysiological recordings can start as soon as 12 h after transfection, depending on ion channel expression level.

3.3 Preparation of PCL Stock Solutions

After synthesis, PCLs are purified as trifluoroacetate or formate salts (7). Compounds are kept with desiccant in the dark at –20 °C to prevent decomposition. Compounds are dissolved either in DMSO or in H_2O to a final concentration usually ≥200 mM. QAQ is dissolved in water. More lipophilic PCLs like AAQ, BzAQ, or PrAQ are dissolved in dry DMSO. In the case of QAQ, we found that the trifluoroacetate salt was less soluble than the formate salt (20 mM vs. 200 mM in water). Stock solutions are aliquoted and kept with desiccant at –20 or –80 °C. Concentration can be checked by UV–vis spectrometry. For all the PCLs described in this chapter the molar extinction coefficient at 360 nm is ε_{360} = 29,000 M^{-1} cm^{-1}. Azobenzenes are very resistant to photobleaching; it is not required to work in the dark when preparing solutions. Working concentrations are usually below 1 mM, which ensures that DMSO content is below 1% during electrophysiological experiments.

3.4 Whole-Cell Recording

The whole-cell configuration (Fig. 2) is a classically used configuration in electrophysiology and is the first step for a pharmacological characterization of a new PCL. The direct control of the potential across the cell membrane allows insight into the molecular interactions between the PCL and the different conformational states of an ion channel. The protocols described here below are deliberately succinct and written for people familiar with electrophysiological techniques. For an introduction to electrophysiological techniques see ref. 26.

There are three main strategies for applying a PCL onto cells:

3.4.1 Perfusion

Various concentrations of PCL can be perfused onto a cell while K_v current is measured in the whole-cell configuration. Current is monitored under 380 and 500 nm light to determine the PCL dose–response relationship and the kinetics of membrane permeation. An obvious disadvantage of this strategy is the significant

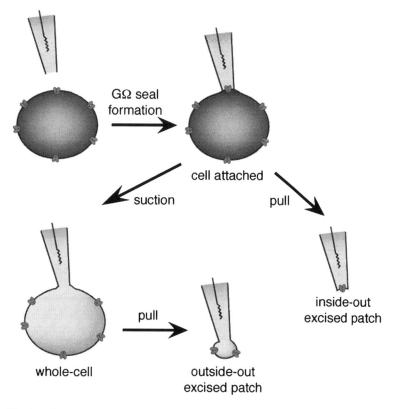

Fig. 2 Different patch-clamp configurations. In a first step, the patch pipette is moved to the surface of the cell membrane and suction is applied to achieve a GΩ seal between the glass pipette and the membrane. This patch configuration is called cell-attached and is used to monitor very high current resolution through single channels. The cell content is intact, making it tricky to determine which ions contribute to the current since the ion concentrations inside the cell are unknown. Cell-attached is also an intermediate step towards getting whole-cell or excised-patch configurations. Inside-out, excised-patch configuration is achieved by pulling the pipette away from the cell. A small piece of membrane will stay attached to the pipette, with the cytosolic part facing the bath solution, allowing perfusion of drugs that normally have a cytosolic mode of action. Whole-cell recording is achieved from cell-attached mode after giving gentle suction to break the piece of membrane at the end of the tip. It is used to measure the average current through the entire cell while controlling ion concentrations on both sides of the membrane. Outside-out, excised patch is obtained from whole-cell recording by pulling the pipette away from the cell. The cell membrane will break and will reform as a convex bleb, leaving a little piece of membrane at the end of the pipette tip, with the external side facing the bath. Single-channel recording can be achieved, with the advantage of having control over ionic concentrations of both sides of the membrane. For electrophysiological characterization of K_v channels photochromic blockers, whole-cell and inside-out configurations have been used

amount of PCLs required, especially since permeation through the lipid bilayer can be slow. It is therefore recommended to perfuse PCLs only when compounds are available in large quantities (tens of mg). In cases where compounds are limited, see Subheadings 3.4.2 and 3.4.3.

Step-by-step protocol:

1. Prepare perfusion lines with various PCL concentrations. The potency of our PCLs for K_v channels usually falls between 10 and 1,000 µM.

2. Place a 12 mm coverslip in the recording chamber.

3. Fill the patch pipette with internal solution, place it into the pipette holder and apply positive pressure. Bring the pipette into the solution of the recording chamber; resistance should be 3–5 MΩ.

4. Select a transfected cell with the appropriate fluorescence filter cube (classically GFP).

5. Once a cell is selected, switch from the fluorescence filter cube to the full mirror position (required for photoswitching).

6. Bring the patch pipette close to the cell while monitoring the pipette resistance with the pClamp software. Stop when the resistance increases by 0.2 MΩ or when a dimple can be seen on the cell surface.

7. Remove positive pressure and start applying gentle suction. Pipette resistance should progressively increase.

8. Once GΩ seal is formed, apply a holding potential of –70 mV.

9. Give brief suctions to break the piece of membrane at the tip of the pipette (Fig. 2). Monitor capacitive currents and resistance through pClamp.

10. Once whole-cell mode is obtained, allow pipette solution to diffuse in the cell for 2–3 min while maintaining a –70 mV holding potential across the membrane.

11. Start recording K_v channel current using protocol described in Fig. 3a (looped at 1 Hz).

12. Perfuse bath solution while running the protocol loop. Because charged PCLs must cross the cell membrane before they bind to K_v channels, ion channel block can take a few of minutes to reach steady state (3–15 min in our case, depending on PCL and concentration, see Fig. 3b). Also, because PCLs are trapped inside cells for a long period of time (typically hours), perfusion can be switched to PCL-free external solution after steady-state value is reached (this saves compound). Figure 3c shows a superimposition of K_v current before perfusion and after perfusion under both wavelengths of light. Finally, because the

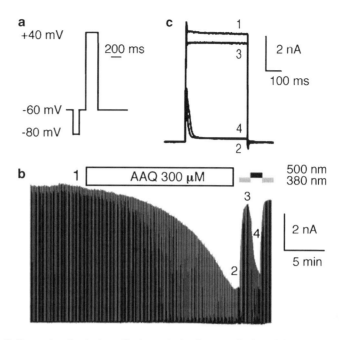

Fig. 3 Example of whole-cell characterization: perfusion. (**a**) Depolarization protocol used to monitor Shaker-IR current. The total protocol length is 1 s. A 100 ms pre-pulse to −80 mV is used to reset K_v channels to their basal state and to correct for eventual leak current. A 200 ms jump to +40 mV is used to open K_v channels. (**b**) Perfusion of 300 µM AAQ on a cell expressing Shaker-IR. K_v current was monitored using the depolarization protocol shown in (**a**) looped at 1 Hz. (**c**) Superimposition of Shaker-IR currents elicited before AAQ perfusion (*1*) and after AAQ perfusion in the dark (*2*) and under 380 (*3*) and 500 (*4*) nm light, at time points shown in (**b**). Capacitive currents have been cut or removed from (**b**) and (**c**) for clarity

compound cannot be washed quickly, it is recommended to start with low concentration and increase concentration until effective dose is determined.

13. Switch on 380 nm light until current is stable.

14. Switch to 500 nm light until current is stable.

3.4.2 Pretreatment

Lipophilic quaternary ammonium can permeate cellular membranes and be trapped inside cells for a prolonged period of time (5, 7, 27). In our case, we found that photosensitivity persisted for several hours after treatment (7). It is therefore possible to pretreat the cell with a certain concentration of PCL, and subsequently record current in whole-cell mode. This strategy has the advantage using much less PCL.

Step-by-step protocol:

1. Remove HEK medium and replace with 500 µl of external solution (Subheading 2.6, step 1) containing PCL (typically

between 10 and 1,000 µM). Incubate for 15 min in the cell incubator (dark, 37 °C).

2. Rinse the cells by placing the coverslip into a 35 mm cell-culture dish containing PCL-free external solution for 2–3 min at room temperature.

3. Place coverslip in recording chamber, obtain a GΩ seal, and go to whole-cell mode as described in Subheading 3.4.1.

4. Record K_v channel current using protocol described in Fig. 3a (looped at 1 Hz) while switching between 380 and 500 nm light.

3.4.3 Loading Through the Patch Pipette

The PCLs we have developed are internal blockers for K_v channels. Hence they can be loaded directly into cells via the patch pipette in whole-cell mode. This strategy, like the PCL pretreatment (Subheading 3.4.2), uses very little compound.
Step-by-step protocol:

1. Dilute the PCL into internal solution (typically 10–1,000 µM, see Subheading 2.5, step 1). The solution must be filtered before inclusion in the patch pipette to remove non-soluble particles. This can be done by placing a syringe filter (0.2 µm) between the syringe and the syringe needle Microfil.

2. Place a 12 mm coverslip in the recording chamber.

3. Obtain a GΩ seal and go to whole-cell mode as described in Subheading 3.4.1.

4. Wait 3–4 min for the PCL to equilibrate in the cytoplasm.

5. Record K_v channel current using protocol described in Fig. 4a (looped at 1 Hz) while switching between 380 and 500 nm light (Fig. 4b, c).

6. Because our PCLs are charged molecules, their binding to K_v channels is expected to be sensitive to membrane potential. Influence of membrane potential on PCL block and unblock can be analyzed using an I/V protocol (Fig. 4d–f).

3.5 Inside-Out Recording

The inside-out configuration (Fig. 2) is used to perfuse drugs to the cytosolic part of ion channels and to study how cytosolic events affect ion channel function. Because our PCLs are internal blockers of K_v channels, the inside-out configuration is used to characterize the IC_{50} (concentration that blocks half of the K_v channel current) of blockers independently of their partition coefficient into the cell membrane.
Step-by-step protocol:

1. Prepare all perfusion lines with various PCL concentrations. Ideally seven solutions are needed for a full IC_{50} determination,

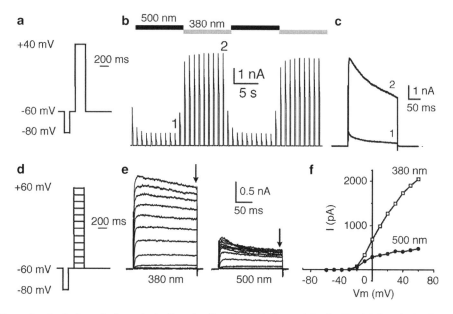

Fig. 4 Example of whole-cell characterization: loading through the patch pipette. (**a**) Depolarization protocol used to monitor Shaker-IR current. The total protocol length is 1 s. A 100 ms pre-pulse to −80 mV is used to reset K_v channels to their basal state and to correct for eventual leak current. A 200 ms jump to +40 mV is used to open K_v channels. (**b**) Shaker-IR current elicited using the protocol described in (**a**). 100 µM QAQ was included in the patch pipette. The protocol was looped at 1 Hz and light was switched every 10 s between 500 and 380 nm. QAQ blocks Shaker-IR current in the *trans* configuration and unblocks it in the *cis* configuration. (**c**) Superimposition of Shaker-IR currents elicited under 500 and 380 nm light, at time points *1* and *2* shown in (**b**), showing block and unblock of QAQ, respectively. (**d**) Depolarization protocol used to monitor Shaker-IR current at different membrane potentials (*I/V* protocol). The total protocol length is 1 s. A 100 ms pre-pulse to −80 mV is used to reset K_v channels to their basal state and to correct for eventual leak current. A 200 ms jump from −60 mV to values between −50 and +60 mV is used to open K_v channels. (**e**) Current through Shaker-IR channels elicited using the *I/V* protocol described in (**d**) and under 380 (*left*) or 500 nm (*right*) light irradiation, using 100 µM QAQ in the patch pipette. (**f**) Steady-state current (at the end of the depolarization, shown by the *arrows* in (**e**)) is plotted as function of membrane potential (*I/V* relationship of block and unblock). Capacitive currents have been cut or removed from (**b**) to (**e**) for clarity

one at the expected IC_{50}, three higher, and three lower. All solutions should be filtered to prevent small debris from breaking the patch.

2. Break the 12 mm coverslip in multiple little pieces (a few mm in diameter) and place one piece into a small recording chamber. The volume of the chamber should be as small as possible, typically 100 µl.

3. Obtain a GΩ seal as described in Subheading 3.4.1, using pipette and bath solutions adequate for inside-out patches (see Subheading 2.6, step 2).

4. Once GΩ seal is formed, apply a positive voltage of +70 mV to set the membrane potential to −70 mV (see Note 6).

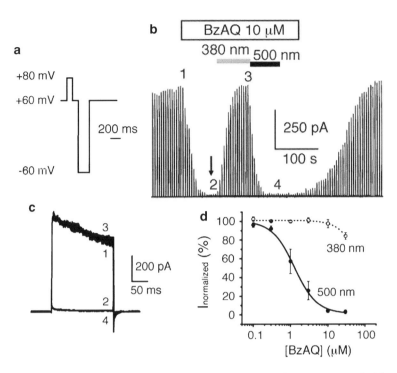

Fig. 5 Example of inside-out characterization using Shaker-IR and BzAQ. (**a**) Command protocol used to open K_v channels in the inside-out configuration (see Note 6). (**b**) In inside-out configuration, K_v current is inward (from the bath to the pipette) but is displayed here as outward. The white bar indicates perfusion of 10 μM BzAQ onto the patch. Arrow indicates when the perfusion is stopped. 380 and 500 nm light irradiations are shown. (**c**) Superimposition of Shaker-IR currents elicited before BzAQ perfusion and during perfusion in the dark and under 380 and 500 nm light illumination, at time points *1, 2, 3,* and *4* shown in (**b**). (**d**) Dose–response of BzAQ block under 380 and 500 nm light. Data points were fitted using the equation: $y = ((A1 - A2)/(1 + ((x/IC_{50})n))) + A2$, where A1 is the maximal current (without PCL), A2 is the current under maximal block, and n is the Hill coefficient. Capacitive currents have been cut or removed from (**b**) and (**c**) for clarity

5. Pull the pipette away from the cell as quickly as possible, using the coarse manipulator.

6. At this stage it is possible that a closed vesicle has formed instead of an open membrane patch (see Note 7). An I/V protocol (similar to Fig. 4d, but see Note 6) can be used to distinguish between a vesicle (no current) and an inside-out patch (typical I/V dependency, like in Fig. 4e, f). Once inside-out patches are obtained, they are very stable over tens of minutes.

7. Record K_v channel current using protocol described in Fig. 5a (looped at 1 Hz).

8. Start perfusing bath solution and wait until K_v current is stable (see number 1 in Fig. 5b).

9. Switch from bath solution to PCL-containing solution, starting from low-concentrated solution. Figure 5b shows K_v current during perfusion of a 10 μM BzAQ solution onto an inside-out patch, which affords full ion channel block.

10. Wait until current is stable and stop the flow (see arrow in Fig. 5b). It is necessary to stop perfusion before switching between different wavelengths of light because photoisomerization rate may be slower than flow rate. Because the chamber has a tiny volume, PCL concentration is theoretically uniform in the entire chamber.

11. Switch on 380 nm light until current is stable (number 3 in Fig. 5b).

12. Switch to 500 nm light until current is stable (number 4 in Fig. 5b).

13. Turn off the light and wash out PCL by perfusing bath solution.

14. Repeat steps 8–13 using a higher PCL concentration.

15. K_v currents can be superimposed to show the extent of block and unblock under the different wavelengths of light (Fig. 5c). The normalized steady-state current can be then plotted as a function of PCL concentration, to determine the IC_{50} for both wavelengths of light (Fig. 5d).

4 Notes

1. Temperature inside the faraday cage can increase substantially when the cage is covered with a curtain. It is recommended to cut a vent in the top of the cage.

2. Polychrome V (Till Photonics) is by default not equipped with a shutter. Till Photonics can add a shutter to the polychrome upon request. It is also possible to use 700 nm light as "dark" since azobenzenes usually do not absorb light beyond 550 nm.

3. The photoisomerization process of azobenzenes is extremely rapid (ps) and population changes can be achieved on a sub-millisecond time scale. Both kinetics and extent of photoswitching are directly correlated to light intensity. Therefore, the more the better: A powerful light source will enable fast photoswitching and maximal photoconversion. We usually measure light intensity through a 20 or 40× objective by hand-placing the power meter at the focal area. Depending on the configuration of the rig (light source, microscope objective, optic fiber, etc.) and the wavelength of light (380 or 500 nm),

light output lies between 0.3 and 5 mW/cm², which is equivalent to $0.5-12.6 \times 10^{15}$ photons/s/cm². When measured through a 20× objective and normalized to the focal area at the specimen plane (estimated to be 300 μm in diameter), light output is between 4 and 70 mW/mm², which is equivalent to $7-176 \times 10^{15}$ photons/s/cm². Converting radiometric unit (mW) to photometric unit (photons/s) can be done using the following equation: $1\ W = \lambda \times 5.03 \times 10^{15}$ photons/s, with λ expressed in nm.

4. 380 nm is a near-UV wavelength of light. It is usually not damaging to tissues. If any concern arises, 400 nm visible light can be used as an alternative, although accumulation of *cis* may be weaker. Tuning of the spectral properties of azobenzene can also be considered (10, 28).

5. *Cis* binders are attractive because in the dark only the "inactive" *trans* form exists (thermodynamically stable); binding to the protein occurs only upon light exposure and the systems returns spontaneously to the baseline in the dark (7, 10).

6. In inside-out configuration, the membrane surface that was originally intracellular is now exposed to bath solution (see Fig. 2). By definition, the patch-clamp command voltage is positive if it increases the potential inside the pipette. As a consequence, a positive command in inside-out mode will depolarize the cell membrane.

7. Vesicles can form inside-out patches when briefly exposed to air, though in our hands this method has a low success rate.

Acknowledgments

We are grateful to Matthew R. Banghart (Harvard Medical School), Michael Kienzler, and Dirk Trauner (University of Munich) for the design and synthesis of PCLs described in this chapter, and to Christopher Davenport for helpful comments and suggestions.

References

1. Kramer RH, Fortin D, Trauner D (2009) New photochemical tools for controlling neuronal activity. Curr Opin Neurobiol 19: 544–552

2. Gorostiza P, Isacoff EY (2007) Optical switches and triggers for the manipulation of ion channels and pores. Mol Biosyst 3:686–704

3. Doyle DA, Morais Cabral J, Pfuetzner RA, Kuo A, Gulbis JM, Cohen SL, Chait BT, Mackinnon R (1998) The structure of the potassium channel: molecular basis of K+ conduction and selectivity. Science 280:69–77

4. Hille B (2001) Ion channels of excitable membranes, 3rd edn. Sinauer Associates Inc, Sunderland, MA, p 814

5. Wang GK, Quan C, Vladimirov M, Mok WM, Thalhammer JG (1995) Quaternary ammonium derivative of lidocaine as a long-acting local anesthetic. Anesthesiology 83:1293–1301

6. Taglialatela M, Vandongen AM, Drewe JA, Joho RH, Brown AM, Kirsch GE (1991) Patterns of internal and external tetraethylammonium block in four homologous K+ channels. Mol Pharmacol 40:299–307

7. Banghart MR, Mourot A, Fortin DL, Yao JZ, Kramer RH, Trauner D (2009) Photochromic blockers of voltage-gated potassium channels. Angew Chem Int Ed 48:9097–9101

8. Fortin D, Banghart M, Dunn TW, Borges K, Wagenaar DA, Gaudry Q, Karakossian MH, Otis TS, Kristan WB, Trauner D, Kramer RH (2008) Photochemical control of endogenous ion channels and cellular excitability. Nat Methods 5:331–338

9. Mourot A, Fehrentz T, Lefeuvre Y, Smith C, Herold C, Dalkara D, Nagy F, Trauner D, Kramer R. Rapid optical control of nociception with an ion channel photoswitch. Nat Methods 9:396–402

10. Mourot A, Kienzler MA, Banghart MR, Fehrentz T, Huber FME, Stein M, Kramer RH, Trauner D (2011) Tuning photochromic ion channel blockers. ACS Chem Neurosci 2:536–543

11. Bartels E, Wassermann NH, Erlanger BF (1971) Photochromic activators of the acetyl-choline receptor. Proc Natl Acad Sci U S A 68:1820–1823

12. Lester HA, Krouse ME, Nass MM, Wassermann NH, Erlanger BF (1979) Light-activated drug confirms a mechanism of ion channel blockade. Nature 280:509–510

13. Volgraf M, Gorostiza P, Szobota S, Helix MR, Isacoff EY, Trauner D (2007) Reversibly caged glutamate: a photochromic agonist of ionotropic glutamate receptors. J Am Chem Soc 129:260–261

14. Nargeot J, Lester HA, Birdsall NJ, Stockton J, Wassermann NH, Erlanger BF (1982) A photoisomerizable muscarinic antagonist. Studies of binding and of conductance relaxations in frog heart. J Gen Physiol 79:657–678

15. Kaufman H, Vratsanos SM, Erlanger BF (1968) Photoregulation of an enzymic process by means of a light-sensitive ligand. Science 162:1487–1489

16. Wainberg MA, Erlanger BF (1971) Investigation of the active center of trypsin using photochromic substrates. Biochemistry 10:3816–3819

17. Westmark PR, Kelly JP, Smith BD (1993) Photoregulation of enzyme activity. Photochromic, transition-state-analogue inhibitors of cysteine and serine proteases. J Am Chem Soc 115(9):3416–3419

18. Zhang Y, Erdmann F, Fischer G (2009) Augmented photoswitching modulates immune signaling. Nat Chem Biol 5:724–726

19. Jiang B, Sun X, Cao K, Wang R (2002) Endogenous K_v channels in human embryonic kidney (HEK-293) cells. Mol Cell Biochem 238:69–79

20. Smart T, Krishek B (1995) Xenopus oocyte microinjection and Ion-channel expression. In: Boulton AA, Baker GB, Walz W (eds) Patch clamp applications and protocols, neuromethods, vol 26. Humana Press, Totowa, NJ, pp 1–47

21. Papazian DM, Schwarz TL, Tempel BL, Jan YN, Jan LY (1987) Cloning of genomic and complementary DNA from shaker, a putative potassium channel gene from drosophila. Science 237:749–753

22. Hoshi T, Zagotta WN, Aldrich RW (1990) Biophysical and molecular mechanisms of Shaker potassium channel inactivation. Science 250:533–538

23. Krouse ME, Lester HA, Wassermann NH, Erlanger BF (1985) Rates and equilibria for a photoisomerizable antagonist at the acetylcholine receptor of electrophorus electroplaques. J Gen Physiol 86:235–256

24. Chen CT, Wagner H, Still WC (1998) Fluorescent, sequence-selective peptide detection by synthetic small molecules. Science 279:851–853

25. Kingston RE, Chen CA, Okayama H. (2003). Calcium phosphate transfection. Curr Protoc Cell Biol, Chapter 20, Unit 20.3

26. Molleman A (2002) Patch clamping: an introductory guide to patch clamp electrophysiology, 1st edn. Wiley, New York, p 186

27. Kirsch GE, Taglialatela M, Brown AM (1991) Internal and external TEA block in single cloned K+ channels. Am J Physiol 261: C583–C590

28. Sadovski O, Beharry AA, Zhang F, Woolley GA (2009) Spectral tuning of azobenzene photoswitches for biological applications. Angew Chem Int Ed 48:1484–1486

Chapter 8

A ¹H NMR Assay for Measuring the Photostationary States of Photoswitchable Ligands

Matthew R. Banghart and Dirk Trauner

Abstract

Incorporation of photoisomerizable chromophores into small molecule ligands represents a general approach for reversibly controlling protein function with light. Illumination at different wavelengths produces photostationary states (PSSs) consisting of different ratios of photoisomers. Thus optimal implementation of photoswitchable ligands requires knowledge of their wavelength sensitivity. Using an azobenzene-based ion channel blocker as an example, this protocol describes a ¹H NMR assay that can be used to precisely determine the isomeric content of photostationary states (PSSs) as a function of illumination wavelength. Samples of the photoswitchable ligand are dissolved in deuterated water and analyzed by UV/VIS spectroscopy to identify the range of illumination wavelengths that produce PSSs. The PSSs produced by these wavelengths are quantified using ¹H NMR spectroscopy under continuous irradiation through a monochromator-coupled fiber-optic cable. Because aromatic protons of azobenzene *trans* and *cis* isomers exhibit sufficiently different chemical shifts, their relative abundances at each PSS can be readily determined by peak integration. Constant illumination during spectrum acquisition is essential to accurately determine PSSs from molecules that thermally relax on the timescale of minutes or faster. This general protocol can be readily applied to any photoswitch that exhibits distinct ¹H NMR signals in each photoisomeric state.

Key words Azobenzene, Photoswitch, Photoisomerization, Photostationary state, NMR

1 Introduction

1.1 Synthetic Photoswitches for Optical Control of Protein Function

The use of light to control protein activity offers distinct advantages over traditional pharmacological manipulations in terms of kinetics and spatial resolution (1, 2). Furthermore, optical control can be combined with genetic targeting to achieve cell-specific manipulations of protein function (3, 4). Two powerful approaches that have been shown to function in living cells and even behaving organisms are based on the use of photochromic ligands (PCLs) and photoswitchable tethered ligands (PTLs) that contain azobenzene photoswitches, which can be reversibly isomerized between *trans* and *cis* isomers with light (4, 5). PCLs are freely diffusing

Matthew R. Banghart (ed.), *Chemical Neurobiology: Methods and Protocols*, Methods in Molecular Biology, vol. 995, DOI 10.1007/978-1-62703-345-9_8, © Springer Science+Business Media New York 2013

small molecule ligands that incorporate the photoswitch at a position crucial to activity at the target receptor. In this configuration, one isomer is more efficacious than the other such that photoisomerization rapidly changes the concentration of the more active isomer to regulate protein activity. PTLs, on the other hand, are covalently attached to the receptor in a way that allows the photoswitch to serve as a tether whose length and geometry can be regulated with light. One end of the photoswitch is functionalized with a protein-reactive moiety, typically an electrophile, and the other end with a ligand (agonist, antagonist, or pore blocker). By attaching the PTL at an appropriate location on the protein surface near the ligand binding site (typically at a mutant cysteine residue), one photoisomer presents the ligand to its binding site at a very high local concentration to regulate the receptor, while the other is geometrically unable to do so and thus inactive. In other related approaches, photoswitches are incorporated into proteins in vitro in a way that allows their conformation to be controlled with light (6, 7).

To understand the utility and limitations of photoswitchable reagents in a biological context, it is imperative to define their photochemical responses to optical stimuli. Ideally, the inactive state of the photoswitch would have no residual activity at the protein target and the active state would lead to complete regulation, or vice versa. Photoswitchable reagents provide the additional advantage of "analogue" or "graded" control over protein activity, as most wavelengths of illumination produce mixtures of isomers. In this way the amount of protein activity can be tuned by varying the illumination wavelength. In practice, it is rarely possible to achieve complete conversion between isomers with light, but it is important to determine these parameters in order to take full advantage of the dynamic range and accurately interpret biological data obtained with photoswitchable tools. Importantly, the wavelength sensitivity and extent of photoconversion must be determined empirically in a context that most closely mimics the cellular environment in which the ligand will be employed. Because many photoswitches, including azobenzenes, thermally relax from the higher energy *cis* isomer back to the low-energy *trans* isomer (8, 9), and do so rather quickly in aqueous solvents in particular (10, 11), continuous illumination is typically required to observe the true PSS (12). Using UV/VIS and NMR spectroscopy at room temperature, this includes photoswitches that thermally relax on the timescale of tens of minutes and faster.

This protocol describes a systematic assay for identifying illumination wavelengths that provide the maximal range of optical control and for quantifying the limits of this range in vitro, using a PTL that photoregulates voltage-gated K+ channels as an example (13) (Fig. 1a). The PTL is composed of a maleimide (MAL) for attachment to cysteine mutant K+ channels, an azobenzene linker (AZO), and a quaternary ammonium ion (QA) that serves as an

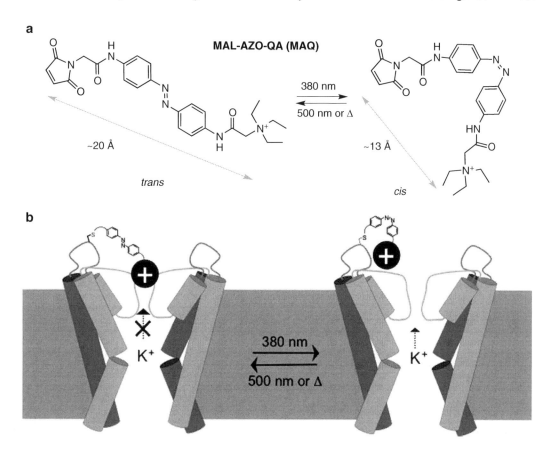

Fig. 1 Optical control of K+ channels with an azobenzene-based photoswitchable tethered ligand (PTL). (**a**) Chemical structure of **MAQ**, which consists of a maleimide for attachment to a mutant cysteine residue on the channel surface, an azobenzene-based photoswitchable linker, and a quaternary ammonium pore blocker. Photoisomerization changes the distance between the tethered end and the pore blocker in a reversible manner. (**b**) Schematic depicting optical control of a K+ channel with **MAQ**. The maleimide component of **MAQ** is attached to a mutant cysteine residue on the channel surface. The position of the cysteine residue and length of **MAQ** are optimized such that the *trans* isomer blocks K+ conductance but the *cis* isomer is too short to reach the pore

extracellular blocker of the ion channel pore, and is thus called **MAQ**. In the *trans* form, the distance between the maleimide and quaternary ammonium ion is ~20 Å, while in the *cis* isomer this distance is significantly shorter at ~13 Å. When attached to a cysteine residue on the channel surface that is ~20 Å from the pore, this molecule blocks conductance (Fig. 1b). However, after photoconversion to the *cis* isomer, the QA is not able to reach the pore, thus restoring ionic conductance. In this way photoisomerization between isomers can be used to reversibly regulate K+ channel function. Although thermal relaxation in the dark (Δ) also allows conversion back to the *trans* isomer, this process takes tens of minutes, whereas photoswitching is complete in seconds or less.

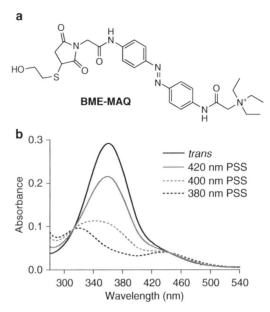

Fig. 2 Photoisomerization of **BME-MAQ** measured by UV/VIS spectroscopy. (**a**) Chemical structure of **BME-MAQ**, an analogue of **MAQ** that has been modified with a water-soluble thiol to mimic bioconjugation to a K+ channel surface. (**b**) UV/VIS spectra of 10 μM **BME-MAQ** in pH 7 PBS buffer in the dark-adapted *trans* state and at various PSSs measured under constant illumination at the indicated wavelengths

To ensure conditions that closely mimic the bioconjugated molecule, we studied a derivative of **MAQ** in which the maleimide has been reacted with β-mercaptoethanol (BME), a water-soluble thiol (**BME-MAQ**, Fig. 2a), and all measurements were taken at neutral pH in aqueous solution. Using the methods described in this protocol, it was determined that 380 nm illumination provides maximal conversion to the *cis* form, and that the 380 nm photostationary state (PSS) under continuous illumination is composed of ~95% *cis*-**BME-MAQ**. On the other hand, 500 nm was identified as the optimal wavelength for conversion to *trans*, producing a PSS of ~80% *trans*-**BME-MAQ**. The implications of these parameters on the use of **MAQ** to photoregulate K+ channels are discussed.

2   Materials

2.1 Sample Preparation

1. Water-soluble azobenzene analogue (i.e., **BME-MAQ**).

2. Deuterium oxide; D, 99.9% (D_2O, Cambridge Isotope Laboratories).

3. Sodium deuteroxide; D, 99.95%, 40% in D_2O (NaOD, Cambridge Isotope Laboratories).

4. Deuterium chloride 100%; D, 99.96%, 20% in D_2O (DCl, Cambridge Isotope Laboratories).

5. 2 mL amber screw-cap vials (VWR).

6. pH paper (VWR).

7. Aluminum foil (VWR).

2.2 UV/VIS Spectroscopy with Continuous Illumination

1. UV/VIS sample: 10 μM azobenzene analogue in D_2O, pH 7.

2. UV-rated cuvette, 50–100 microliter volume with a 10 mm path length (VWR).

3. Hewlett Packard 8453 UV/VIS spectrometer (HP). If using another model, it must function with an open top for fiber-optic access.

4. Polychrome V monochromator containing a 150 W Xenon short arc lamp with an output range of 320–680 nm and a half-power bandwidth of 14 nm (Till Photonics, see Note 1).

5. Factory fiber optic cable (Till Photonics).

6. Small cart for the monochromator (Staples).

7. Narrow diameter two-prong extension clamp (VWR).

8. 12″ aluminum rod (VWR).

9. Support stand (VWR).

10. Safety goggles rated to protect against the wavelength and light intensity range of the monochromator (Thorlabs).

2.3 ^{1}H NMR with Continuous Illumination

1. NMR sample: 100 μM azobenzene analogue in D_2O, pH 7.

2. 500 MHz, 5 mm, 7 in. long screw-capped NMR tube (Wilmad Lab Glass).

3. FT-600-UMT fiber optic cable (at least 20 ft, Thorlabs) fitted with an adaptor to the monochromator on one end and left bare on the other (see Note 2).

4. Bruker AV500 NMR spectrometer operating at 500.23 MHz (^{1}H) equipped with a triple resonance TBI [^{1}H, ^{13}C, X] probe-head (Bruker) (see Note 3).

5. 7 in. long 22 gauge needle (VWR).

6. 1 mL syringe (VWR).

3 Methods

3.1 Sample Preparation

1. A small quantity (several milligrams or less) of dry sample is weighed out into a screw-cap vial (see Note 4).

2. A stock is prepared by dissolving the sample in D_2O in the millimolar concentration range (see Note 5).

3. An aliquot of the stock is diluted to produce a 1 mL NMR sample at a concentration of 100 µM in D_2O (see Note 5).

4. The NMR sample pH is checked by placing several microliters of the sample on pH paper. If the pH is not ~7 (or the pH range of the intended biological environment), it is carefully adjusted using DCl or NaOD without largely reducing the concentration of analyte (see Note 6).

5. A 10 µl aliquot of the 100 µM sample is further diluted to produce a 10 µM UV/VIS sample in D_2O (see Note 5).

6. The UV/VIS sample pH is checked again on pH paper and adjusted if needed.

3.2   UV/VIS with Continuous Illumination

1. The monochromator is fitted with the factory fiber-optic cable, turned on, and allowed to warm up for ~15 min prior to use. Safety goggles should be worn while the light source is turned on.

2. Under dim room light, a minimal volume (e.g., 50 µl) of D_2O is added to the cuvette and a "blank" spectrum is obtained.

3. The D_2O is replaced with the same volume of dark-adapted UV/VIS sample.

4. A UV/VIS spectrum of the *trans* isomer is obtained across the relevant wavelength range (typically 280–600 nm). The spectrum of *trans*-**BME-MAQ** is shown in Fig. 2b.

5. With illumination disabled, the output collar of the monochromator-coupled fiber-optic cable is mounted above the cuvette using the clamp/rod/base assembly. If the spectrometer is equipped with a stir plate, the sample is stirred. The tip of the cable is placed several millimeters above the liquid to ensure complete illumination of the sample with high-intensity light. The Polychrome V does not contain a shutter. To disable illumination, the fiber is disconnected and aluminum foil is placed over the monochromator port to block the emitted light.

6. The monochromator is tuned to a wavelength that is strongly absorbed by the *trans* isomer and thus predicted to cause substantial photoisomerization (typically 360–460 nm) and illumination is enabled. After 10–30 s of illumination, a spectrum is obtained without disengaging the light source. The spectrum should resemble that of the *trans* isomer but with reduced intensity and possibly contain additional peaks, representing a mixture of *trans* and *cis* isomers, as is observed for **BME-MAQ** in Fig. 2b (see Note 7).

7. After an additional 10–30 s of illumination, another spectrum is obtained. If the spectrum is different from the previous spectrum, further illumination is required to be certain that the PSS has been reached. This process is repeated until the spectrum does not change further. The stable spectrum represents the PSS produced by the selected illumination wavelength.

8. The monochromator is tuned to another wavelength (20 nm increments are convenient) and spectra are obtained until the PSS is reached.

9. Step 8 is repeated until the range of wavelengths that produce PSSs have been covered in 20 nm increments. The spectra of **BME**-**MAQ** at the PSSs produced by illumination at 380 nm, which contains predominantly *cis* isomer, 400 nm, which contains a mixture enriched in *cis* isomer, and 420 nm, which contains a mixture enriched in *trans* isomer, are shown in Fig. 2b.

10. To determine the precise wavelengths that cause maximal photoconversion between isomers, step 8 is repeated at smaller wavelength increments. The spectrum of the PSS at the optimal wavelength for producing *trans* isomer will most closely resemble the dark-adapted *trans* spectrum. The spectrum of the PSS at the optimal wavelength for producing *cis* isomer will contain the lowest intensity *trans* isomer peak. *Trans*-**BME-MAQ** is barely detectable in the 380 nm PSS (Fig. 2b) but not every azobenzene analogue will yield the same degree of photoconversion.

3.3 ¹H NMR with Continuous Illumination

1. The monochromator is fitted with the FT-600-UMT fiber-optic cable, placed near the NMR spectrometer, but outside the magnetic field, turned on, and allowed to warm up for ~15 min prior to use. Safety goggles should be worn when the light source is turned on.

2. Under dim room light, 700 µl of the dark-adapted NMR sample is added to the bottom of the screw-capped NMR tube using the 1 mL syringe and long needle. In doing so, the needle punctures the rubber top of the screw cap, providing a port for the fiber-optic cable (see Note 5).

3. The NMR tube containing the sample is placed in the NMR sample holder provided with the instrument and adjusted to the appropriate depth in the holder.

4. With illumination disabled, the fiber-optic cable is inserted through the rubber screw cap and lowered into the tube so that it rests ~1 mm above the liquid surface. The rubber cap should hold tightly to the cable so that the NMR tube/holder assembly can be suspended from the cable (Fig. 3a).

5. With the sample lift turned off, the NMR tube/holder assembly is carefully lowered into the NMR instrument using the fiber-optic cable until the sample holder is in place, as indicated by a lack of pull on the cable by gravity (see Note 8). Care should be taken to ensure that the fiber-optic cable does not provide any lateral pressure on the assembly that might cause the NMR tube to sit at an angle in the spectrometer. This can be achieved by securing a portion of the cable to part of the spectrometer housing using a string or by other means.

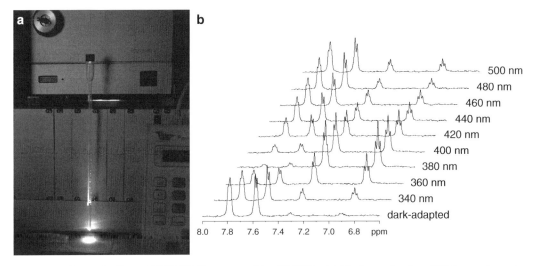

Fig. 3 Photoisomerization of **BME**-**MAQ** measured by ^{1}H NMR. (**a**) Photograph of an NMR sample under constant illumination via a fiber-optic cable connected to the Polychrome V monochromator. (**b**) Aromatic ^{1}H NMR spectra of **BME**-**MAQ** in pH 7 D$_2$O in the dark-adapted state and at various PSSs measured under constant illumination at the indicated wavelengths

6. The NMR spectrometer is tuned to maximize sensitivity in the ^{1}H channel. Consult with the NMR facility specialist for assistance with this step.

7. A ^{1}H NMR spectrum of the *trans* isomer is acquired over 6 kHz centered at 4.5 ppm into 32 K data points (acquisition time = 1.6 s). To attain a reasonable signal-to-noise ratio, 512 scans are typically required using a TBI probe, resulting in a total acquisition time of approximately 15 min. Additional scans can be added to obtain a suitable signal-to-noise ratio. The aromatic region (8.0–6.6 ppm) of the ^{1}H NMR spectrum of dark-adapted *trans*-**BME**-**MAQ** is shown in Fig. 3b (see Note 9).

8. The monochromator is tuned to a wavelength previously determined to produce a PSS enriched in the *cis* isomer and illumination is enabled. After 10–30 min, a ^{1}H NMR spectrum is obtained using the acquisition parameters determined in step 7. The spectrum should contain peaks corresponding to the *trans* isomer obtained in step 7 but also additional peaks that correspond to the *cis* isomer as is observed in Fig. 3b. The *cis* azobenzene aromatic protons typically resonate upfield from the *trans* peaks (see Note 7).

9. After an additional 10–30 min of illumination, another spectrum is obtained. If the spectrum is different from the previous spectrum, further illumination is required to ensure that the PSS has been reached. This process is repeated until the spectrum does not change further.

10. The monochromator is tuned to another wavelength and spectra are obtained until the PSS is reached. The spectra of **BME-MAQ** at the PSSs produced by illumination at 20 nm increments from 340 to 500 nm are shown in the aromatic region in Fig. 3b.

11. Step 10 is repeated until the desired range of wavelengths has been covered.

12. To determine the relative fractions of *cis* and *trans* isomer at each PSS, the signals of distinct proton atoms that are well resolved between isomers are integrated (Int). Because a ratio will be calculated, normalization is not necessary. The % *cis* isomer is determined as the $\mathrm{Int}_{cis} / (\mathrm{Int}_{cis} + \mathrm{Int}_{trans}) \times 100$.

13. This procedure is repeated for three independent NMR samples and the results are averaged for quantification.

4   Analysis and Discussion

For comparison, the spectral sensitivity was also examined in a functional assay using electrophysiological recordings from **MAQ**-treated HEK293 cells expressing a cysteine mutant Shaker K+ channel (13, 14) (Fig. 4a). Currents were evoked using depolarizing voltage steps to open the channels while constant illumination was applied at various wavelengths in 1 min intervals. For clarity and normalization purposes, channels were closed with 500 nm light in between test wavelengths of shorter wavelengths. The same light source described above was coupled to a microscope and focused on the recorded cell. Under these conditions, PSSs were reached within 15–20 s of changing the wavelength, and the currents

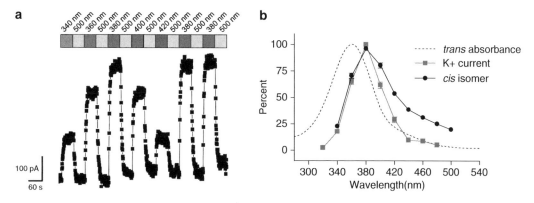

Fig. 4 Functional implications of PSSs. (**a**) Electrophysiological recording of K+ currents through **MAQ**-modified cysteine mutant K+ channels at different illumination wavelengths. (**b**) Superimposition of the normalized *trans*-**BME-MAQ** absorbance spectrum, the % *cis*-**BME-MAQ** measured at each PSS by ¹H NMR, and the normalized action spectrum measured electrophysiologically from **MAQ**-modified K+ channels

measured during the last 30 s of each trial were averaged for each wavelength across several cells. These measurements were normalized to the current values observed at the 380 nm PSS for comparison to the other data.

This action spectrum is plotted in Fig. 4b along with the UV/VIS absorbance spectrum measured for *trans*-**BME**-**MAQ**, and the PSS spectrum measured for **BME**-**MAQ** by ^{1}H NMR. From these comparisons several key features are apparent that provide important insights into how **MAQ** can be used to control K+ channel function.

Examination of UV/VIS spectrum of *trans*-**BME**-**MAQ** alone might suggest that the absorbance peak at 360 nm is the optimal wavelength for conversion to *cis*. However, it is the product of the *trans/cis* absorbance ratio and corresponding ratios of the quantum yields of photoisomerization that determines the PSS at a given wavelength (15, 16). Thus the absorbance minimum of the *cis* isomer is most commonly the optimal wavelength for conversion to *cis*. In general it is very difficult if not impossible to isolate the *cis* isomer of azobenzenes that thermally relax on a timescale faster than several hours. Thus it is difficult to predict the optimal wavelengths without a spectrum of the pure *cis* isomer. The UV/VIS spectra of the PSSs generated at various wavelengths are sufficient to determine the optimal wavelengths for photoconversion between isomers. But this information does not reveal the isomeric content at each PSS. The 4,4′-diamidoazobenzene chromophore used in **MAQ** has the advantageously rare feature that PSSs of >95% *cis* isomer can be obtained (at 380 nm). In this case, the UV/VIS spectrum of the 380 nm PSS can be used as an excellent approximation for the pure *cis* isomer. In less convenient, more typical cases, the isomeric content of at least one PSS must be known or estimated in order to calculate the *cis* isomer spectrum, which can then be used to predict the isomeric content at any PSS. Alternatively, the photokinetic method can be used to calculate the *cis* isomer absorbance spectrum (17).

The use of ^{1}H NMR spectroscopy simplifies matters considerably. By providing a direct measurement of the PSS, estimations are not required. Thus the maximal extent of photoconversion between isomers that can be achieved is readily determined. Because the absorbance spectra of both isomers overlap substantially, it is very difficult to predict the isomeric content using UV/VIS spectra of PSSs, particularly at intermediate wavelengths (see Fig. 2b). Yet these PSSs can be valuable in physiological experiments that capitalize on the analogue control provided by photoswitchable reagents (18).

These NMR data for the model compound **BME**-**MAQ** correlate well with the electrophysiological action spectrum obtained with **MAQ** bound to Shaker channels (Fig. 4b). While the shapes of the curves are somewhat different at longer wavelengths, the fact that both peak around 380 nm indicates that conjugation of **MAQ** to the channel surface in living cells does not dramatically perturb its photochromic profile. Additionally, the NMR data show that at the optimal wavelength (380 nm), the *trans* isomer can be almost completely converted to the *cis* isomer. However, photoconversion back to *trans* isomer is not as complete, with 500 nm producing a PSS of only ~80% *trans*. Taken together, these parameters suggest that although nearly complete unblock of a K+ channel conjugated with a single molecule of **MAQ** should be achieved with 380 nm illumination, only 80% block should be expected, leaving a substantial residual current light-insensitive. In contrast with these predictions, experiments in which **MAQ** is added to channels during the recording instead revealed complete block and incomplete unblock with light (13). The most likely explanation lies in the tetrameric nature of voltage-gated K+ channels such as Shaker. Because all four subunits contain a cysteine residue in these experiments, each channel may contain up to four copies of **MAQ**. This would not only increase the effective concentration of pore blocker under 500 nm illumination, but it would also render the probability of pore occupancy dependent on the fourth power of the PSS. Thus at the PSS composed of 80% *trans*, approximately three of the four **MAQ** molecules per channel would be in the blocking state at any moment in time, leading to >99% block. Similarly, at the 95% *cis* PSS, four copies of **MAQ** would still block ~20% block of the total current through the channels, which is more consistent with the functional data (13), although it is unlikely that each channel is actually modified at all four cysteines under these conditions (19, 20). In future applications of PTLs such as **MAQ**, such a precise knowledge of PSS content can be utilized to optimize the extent of photocontrol over protein function by regulating multivalency through genetic engineering (14, 20).

5 Notes

1. Alternative light sources consisting of a lamp and filters may be suitable. However, Xenon arc lamps provide relatively uniform light power across the UV/VIS spectrum which is important when examining molecules that thermally relax rapidly, and are thus preferred.

2. The fiber adaptor used in this study was custom built and consisted of a ferrule that could be clamped into the center of the monochromator light port. The output of the fiber-optic cable between 340 and 480 nm ranged from 0.3 to 9.0 $\mu W/cm^2$. To achieve greater power, a more sophisticated adaptor would contain a lens to focus the light onto the fiber tip. In theory, higher power can be obtained by using any larger diameter

multimode fiber that transmits the requisite wavelength range, as long as the bare fiber is of small enough diameter to fit inside the 5 mm NMR tube when stripped at one end.

3. The use of high field magnet combined with a TBI [^{1}H, ^{13}C, X] probe provides excellent ^{1}H sensitivity and thus is highly advantageous, as very dilute samples are required to achieve photoswitching at the light power levels readily obtained from the FT-600-UMT cable. If lower ^{1}H sensitivity configurations are to be used, effort should be made to maximize light delivery from the fiber-optic cable.

4. Before use, samples should be thoroughly dried to remove traces of solvents and atmospheric water. This can be achieved using a vacuum pump followed by storage over desiccant or by the most effect means available. Residual solvents and water can reduce the intensity of the sample signal in ^{1}H NMR experiments.

5. Once the sample is dissolved in D_2O, it should be kept tightly closed and protected from the atmosphere as much as possible to prevent water from contaminating the NMR spectrum. The empty volume above the solution can be filled with nitrogen or argon gas that has been passed through desiccant before capping and sealing with parafilm. Furthermore, once the sample is dissolved in solvent, it should be protected from light. Room lights can be dimmed or turned off, and the vials and/or NMR tubes should be covered in aluminum foil.

6. Absorbance properties, photoisomerization rates and presumably quantum yields are pH dependent due to the protonation of the azo-bond nitrogen atoms or other basic substituents on the azobenzene ring (21). This is particularly important when the compound under examination contains acid/base functionality or exists as a salt, as is the case for the tetraalkylammonium group in **BME**-**MAQ**. pH can be adjusted using DCl and NaOD to avoid contamination with H_2O. To avoid overcompensation, these can be pre-diluted with D_2O into working solution aliquots.

7. If no change is observed at a wavelength that is strongly absorbed by the *trans* isomer, several parameters can be optimized. The ability of light to penetrate the sample is determined by the light intensity and the optical density of the sample. The latter depends on the extinction coefficient of the chromophore and sample concentration according to the Beer-Lambert law. First, light power should be optimized, possibly by adjusting the monochromator-to-fiber coupler. The fiber output should be moved as close to the liquid surface of the sample as possible to ensure that the highest intensity light is concentrated on the sample. If a change in the spectrum is still not observed,

the path length through which the light must pass may be too large. The volume of sample should be reduced to the minimal volume that can be assayed by the spectrometer. Similarly, the sample may be too concentrated for the available light intensity to penetrate. The sample should be diluted in half and the process repeated. If the sample is diluted to a point that the sample signal is too weak for the spectrometer to detect, the azobenzene analogue under investigation may not yield appreciable PSSs in aqueous solution at room temperature. This is typically due to thermal isomerization occurring faster than photoexcitation at the available light intensity. If possible, the sample can be cooled to slow thermal relaxation. Alternatively, organic cosolvents such as deuterated methanol or dimethylsulfoxide can be added, as azobenzene thermal isomerization is typically slower in less polar, aprotic solvents than in water (11).

8. Care should be taken to keep the light source and all metal-containing optical components outside the NMR magnetic field.

9. If the sample contains a large amount of contaminating water that limits the ability to detect the azobenzene signal, a water suppression protocol can be applied.

Acknowledgments

The authors would like to thank Jessica Harvey for contributing to the synthesis of **MAQ**, Rudi Nunlist of the UC Berkeley Department of Chemistry NMR facility for assisting with the NMR experiments, and Enrique Chang, formerly at Till Photonics for coordinating customization of the Polychrome V used in these experiments.

References

1. Ellis-Davies GC (2007) Caged compounds: photorelease technology for control of cellular chemistry and physiology. Nat Methods 4(8):619–628

2. Szobota S, Isacoff EY (2010) Optical control of neuronal activity. Annu Rev Biophys 39:329–348

3. Miesenbock G (2011) Optogenetic control of cells and circuits. Annu Rev Cell Dev Biol 27:731–758

4. Fehrentz T et al (2011) Optochemical genetics. Angew Chem Int Ed Engl 50(51): 12156–12182

5. Gorostiza P, Isacoff EY (2008) Optical switches for remote and noninvasive control of cell signaling. Science 322(5900):395–399

6. Beharry AA, Woolley GA (2011) Azobenzene photoswitches for biomolecules. Chem Soc Rev 40(8):4422–4437

7. Banghart MR et al (2006) Engineering light-gated ion channels. Biochemistry 45(51): 15129–15141

8. Rau H (2003) Azo compounds. In: Durr H, Bouas-Laurent H (eds) Photochromism: molecules and systems, revised edition. Elsevier, San Diego, pp 165–192

9. Knoll H (2004) Photoisomerism of azobenzenes. In: Horspool W, Lenci F (eds) CRC handbook of organic photochemistry and photobiology, 2nd edn. CRC Press, Boca Raton, pp 89/1–89/16

10. Hartley GS (1938) Cis form of azobenzene and the velocity of the thermal cis/trans conversion of azobenzene and some derivatives. J Chem Soc 633–642

11. LeFevre RJW, Northcott J (1953) The effects of substituents and solvents on the cis→trans change of azobenzene. J Chem Soc 867–870

12. Tait KM et al (2003) The novel use of NMR spectroscopy with in situ laser irradiation to study azo photoisomerization. J Photochem Photobiol A Chem 154:179–188

13. Banghart M et al (2004) Light-activated ion channels for remote control of neuronal firing. Nat Neurosci 7(12):1381–1386

14. Fortin DL et al (2011) Optogenetic photochemical control of designer K+ channels in mammalian neurons. J Neurophysiol 106(1): 488–496

15. Fischer E et al (1955) Wave length dependence of photoisomerization equilibria in azo compounds. J Chem Phys 23:1367

16. Zimmerman G et al (1958) The photochemical isomerization of azobenzene. J Am Chem Soc 80:3528–3531

17. Borisenko V, Woolley GA (2005) Reversibility of conformational switching in light-sensitive peptides. J Photochem Photobiol A Chem 173(1):21–28

18. Banghart MR et al (2009) Photochromic blockers of voltage-gated potassium channels. Angew Chem Int Ed Engl 48(48): 9097–9101

19. Blaustein RO et al (2000) Tethered blockers as molecular 'tape measures' for a voltage-gated K+ channel. Nat Struct Biol 7(4):309–311

20. Blaustein RO (2002) Kinetics of tethering quaternary ammonium compounds to K(+) channels. J Gen Physiol 120(2):203–216

21. Sawicki E (1957) Physical properties of aminoazobenzene dyes. VIII. Absorption spectra in acid solution. J Org Chem 22:1084–1088

Chapter 9

Developing a Photoreactive Antagonist

Pamela M. England

Abstract

Light is an exquisite reagent for controlling the activity of biological systems, often offering improved temporal and spatial resolution over strictly genetic, biochemical, or pharmacological manipulations. This chapter describes a general approach for developing small molecules that, upon irradiation with light, may be used to rapidly inactivate targeted proteins expressed on the surfaces of cells. Highlighted is ANQX, a photoreactive AMPA receptor antagonist developed to irreversibly inactivate a subtype of glutamate-gated ion channels natively expressed on neurons.

Key words AMPA receptor, Photoreactive antagonist, DNQX, Photoaffinity label, Receptor trafficking

1 Introduction

Signaling in the brain depends on the ability of nerve cells to respond to small molecules. Presynaptic action potentials trigger the release of chemical neurotransmitters into specialized regions called synapses, wherein one neuron receives signals from another. Subsequent binding of the neurotransmitters to ligand-gated ion channels on postsynaptic neurons causes the channels to open, resulting in the flow of ions into the neuron. This electrochemical signaling, termed synaptic transmission, underlies the fast communication between neurons (Fig. 1). Moreover, regulated changes in the strength of synaptic transmission provide the molecular basis for modifications in neuronal function that underlie myriad physiological and pathophysiological processes, including formation of memories and addiction to drugs.

Considerable evidence suggests that changes in the strength of synaptic transmission at many synapses are due to activity-dependent changes in the number of ion channels present on postsynaptic membranes (1–6). Receptors may be inserted or removed from synapses by exocytosis (movement from intracellular stores to the surface of the neuron), endocytosis (movement from the

Matthew R. Banghart (ed.), *Chemical Neurobiology: Methods and Protocols*, Methods in Molecular Biology, vol. 995, DOI 10.1007/978-1-62703-345-9_9, © Springer Science+Business Media New York 2013

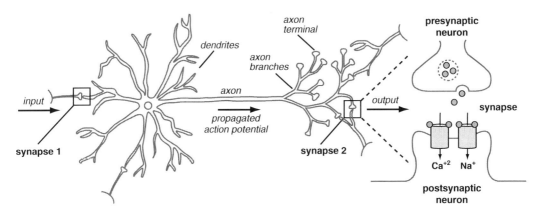

Fig. 1 Synaptic transmission involves the release of neurotransmitter from a presynaptic neuron, into the synaptic cleft, followed by binding of the neurotransmitter to ligand-gated ion channels on a postsynaptic neuron. The flow of ions through the postsynaptic receptor converts the chemical signal back into an electrical one that can be propagated along the postsynaptic neuron to the next synapse

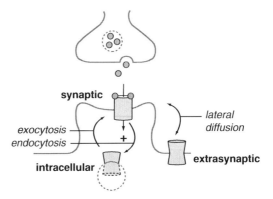

Fig. 2 The trafficking of receptors into and out of synapses involves the exocytosis of receptors from inside of the cell to the membrane surface, the removal of receptors from the membrane surface by endocytosis, and the lateral diffusion of receptors along the membrane surface

surface of the neuron to intracellular sites), and lateral diffusion (movement along the surface of the neuron) (Fig. 2). Delineating the precise trafficking mechanisms that are active under various conditions is a central focus of modern molecular neuroscience.

Light-activated small molecules represent a powerful class of tools for rapidly manipulating and monitoring the trafficking of ligand-gated ion channels. These tools primarily fall into two categories—photocaged and photoreactive ligands. Photocaged ligands, typically caged agonists, are used to rapidly control the activation of target receptors (7–9). These ligands are inert until irradiated with light, whereupon they are converted into fully active ligands. Contrastingly, photoreactive ligands bind to their native receptors before being activated with light and have classically

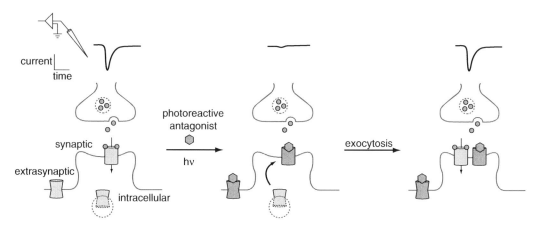

Fig. 3 Monitoring receptor trafficking (exocytosis) using photoinactivation and continuous electrophysiological recordings. Agonist-induced ionic currents recorded from intact receptors prior to application of the photoreactive ligand provide a baseline response. Following application of the photoreactive antagonist in the presence of ultraviolet light, the baseline response is immediately reduced. Agonist-induced ionic currents return as intact receptors from the inside of the cell are exocytosed onto the surface of the cell

been used to identify cognate receptors and the ligand binding sites contained within them (10). Upon irradiation with ultraviolet light these ligands are converted to highly reactive intermediates capable of forming covalent adducts with bound receptors. Photoreactive ligands have recently found another important application, being used to rapidly inactivate ligand-gated ion channels expressed on the surfaces of cells (11–13). Photochemical inactivation, used in combination with electrophysiological recordings of neuronal activity, provides a means of directly monitoring the trafficking of native receptors in real time (Fig. 3). The approach, which my laboratory demonstrated for AMPA receptors, relies on developing a photoreactive antagonist for the channel of interest (11, 14–16). This chapter describes the development of a photoreactive antagonist for AMPA receptors.

2 Designing a Photoreactive Ligand

The strategy used to design a photoreactive antagonist for AMPA receptors is a general one that can be applied to the development of photoreactive ligands for practically any receptor of interest. Three things are necessary to develop a photoreactive ligand: (1) a core structure for the ligand, (2) an appropriate photoreactive moiety to attach to the ligand, and (3) an assay for quantifying the activity of the ligand.

The first step in designing a photoreactive ligand is choosing a core structure, based on an established pharmacology for the receptor. Numerous classes of agonists and antagonists have been developed for most receptors and channels expressed in the nervous system, including AMPA receptors (17). These classes of compounds have largely been developed by the pharmaceutical industry in an effort to identify more potent and/or selective receptor ligands. Within each compound class there may be hundreds of analogs of a single core structure, displaying a range of activities. Structure–activity studies of these analogs reveal which modifications impact binding to the receptor. Thus, if one considers a core structure to be a square with four functional ("R") groups (see Fig. 4), structure–activity studies suggest which "R" groups are and are not important for binding. An appropriate core structure is one that contains an R group that can range in size (e.g., methyl, ethyl, propyl) without significantly impacting binding to the receptor. This R group will be replaced by a photoreactive moiety in developing the photoreactive ligand.

The core structure we selected was the competitive antagonist 6,7-dinitro-quinoxaline-2,3-dione (DNQX) (Fig. 5). Numerous analogs of DNQX have been prepared and based on these structures it was evident that one of the nitro groups on DNQX could be replaced with a variety of moieties without significantly affecting binding to the receptor (11, 17–22).

The second step in designing a photoreactive ligand is choosing the photoreactive functional group. Three photoreactive groups are commonly used in photo-cross-linking studies—the azide, the benzophenone, and the diazarine (Fig. 6) (10). Each group has advantages and disadvantages. The azide and diazarine are irreversibly transformed into radicals upon irradiation with ultraviolet light. These radicals rapidly react with nearest neighboring atoms, whether they are located on the receptor of interest or on water molecules in the surrounding medium. The cross-linking efficiency of these groups is rather low, typically ~1 %. However, the cross-linking efficiency is not an issue in the case of immobilized receptors, such as ligand-gated ion channels expressed on the surfaces of cells, as "fresh" (nonirradiated) ligand can be continuously perfused over the cell and irradiated until photo-cross-linking is complete. The benzophenone functional group reversibly enters a triplet state upon photostimulation with ultraviolet light. Cross-linking occurs following hydrogen atom extraction by the triplet benzophenone from alkyl groups (e.g., protein side-chains) and subsequent coupling of the resulting radicals. The selectivity for alkyl groups (i.e., C–H bonds, but not O–H bonds found in water) may obviate the need to perfuse the receptor with fresh ligand. The disadvantage of benzophenone is its size; such a large substitution is more likely to abrogate binding to the target receptor than the smaller azide or diazarine groups. It should be noted that the precise wavelength of ultraviolet light that the azide, diazarine, and benzophenone moieties responds to will vary depending on the nature of the ligand they are attached to. As a general rule, wavelengths

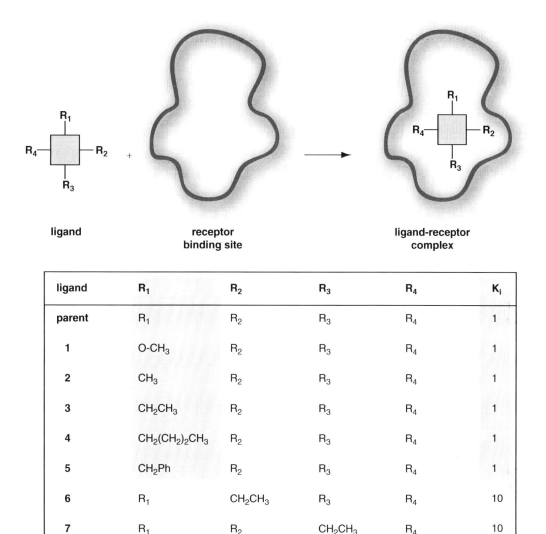

ligand	R_1	R_2	R_3	R_4	K_i
parent	R_1	R_2	R_3	R_4	1
1	O-CH$_3$	R_2	R_3	R_4	1
2	CH$_3$	R_2	R_3	R_4	1
3	CH$_2$CH$_3$	R_2	R_3	R_4	1
4	CH$_2$(CH$_2$)$_2$CH$_3$	R_2	R_3	R_4	1
5	CH$_2$Ph	R_2	R_3	R_4	1
6	R_1	CH$_2$CH$_3$	R_3	R_4	10
7	R_1	R_2	CH$_2$CH$_3$	R_4	10
8	R_1	R_2	R_3	CH$_2$CH$_3$	10

Fig. 4 A hypothetical structure–activity table. The relationship between the functional groups on the ligand (R_1, R_2, R_3, R_4) and the K_i (or other measures of affinity such as IC$_{50}$) reveals which group(s) may be replaced with a photoreactive moiety, without abrogating binding to the target receptor. In the case of the hypothetical ligand in this table, replacing R_1 with a variety of functional groups has no effect on K_i, whereas changing R_2, R_3, or R_4 decreases the affinity of the ligand for the receptor tenfold

less than 320 nm should be avoided as they are typically damaging to cells.

In our case we chose to replace a nitro group on DNQX with an azide to form the photoreactive ligand ANQX (6-azido,7-nitro-quinoxaline-2,3-dione) (Fig. 6). It is worth noting that although a crystal structure of the AMPAR ligand-binding domain was available, this information did not influence the design of ANQX.

Compound	[³H]AMPA (Kᵢ, μM)	Electrophysiology (IC₅₀, μM)	Reference
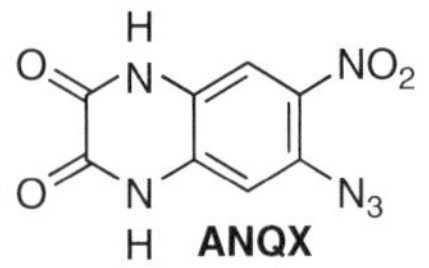 **ANQX**	nd	0.5	11
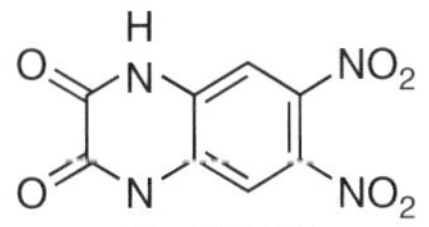 **DNQX**	0.20	1.0	11, 20
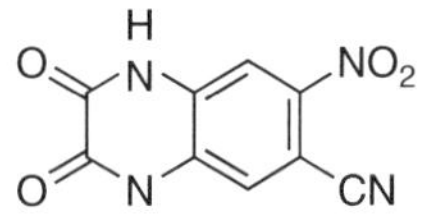 **CNQX**	0.27	0.6	20, 21
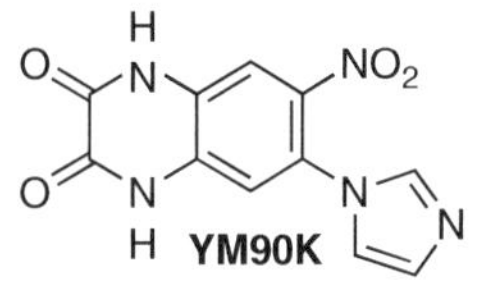 **YM90K**	0.08	0.3	20, 22
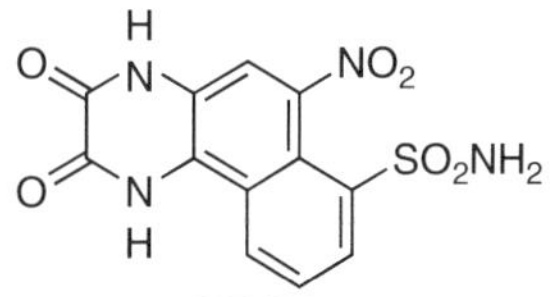 **NBQX**	0.06	0.2	20,21

Fig. 5 Structures and inhibition constants for several structurally related quinoxalinediones. The binding of (³H) AMPA was carried out with crude rat brain membranes and the K_i values were determined using the Cheng–Prusoff relationship. IC₅₀ values were determined by measuring dose–response curves from AMPA receptors heterologously expressed in Xenopus oocytes

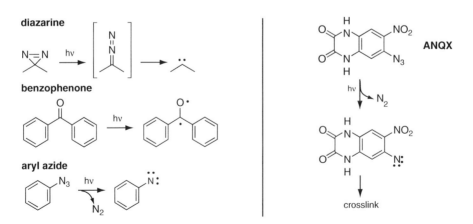

Fig. 6 Structures and photochemistry of common photo-cross-linkers (diazarine, benzophenone, and aryl azide), and the photoreactive AMPA receptor antagonist ANQX

Molecular docking to the ligand-binding domain indicated that ANQX could bind in either of two orientations—with the azide pointing into the ligand-binding pocket or protruding from the pocket. An X-ray crystallography study we subsequently carried out revealed that ANQX bound to the receptor in the former orientation (16).

The final step in designing a photoreactive ligand is to develop an assay for quantifying the affinity of the intact ligand for the receptor as well as the efficacy with which it photo-cross-links to the receptor. In the case of ANQX, we used two-electrode voltage clamp recordings of glutamate-evoked currents from recombinant receptors heterologously expressed in Xenopus oocytes. Electrophysiology is an exquisitely sensitive technique, providing a robust signal from as little as femtomoles of receptor expressed on the surface of an oocyte. An IC_{50} for the intact photoreactive antagonist can be obtained from standard dose–response curves generated by measuring agonist-induced currents in the presence of varying amounts of antagonist. In the case of ANQX, the IC_{50} at AMPARs was 0.5 μM, differing only by a factor of two from the parent compound DNQX ($IC_{50} = 1.0$ μM). The efficacy of cross-linking is quantified by monitoring the extent of the decrease in agonist-induced currents per unit time of photostimulation in the presence of the photoreactive antagonist. If the target receptor is not itself electrophysiologically "active," indirect methods for determining the efficacy of cross-linking may be available. For example, in the case of G-protein-coupled receptors, co-expression of the receptor with a G-protein-coupled inwardly rectifying potassium (GIRK) channel can be used to electrophysiologically monitor the activity of the receptor (23).

A number of controls are required to demonstrate that the observed effects of photostimulation are directly due to cross-linking of the photoreactive ligand to the target receptor. First, to demonstrate that ultraviolet light alone has no effect on either the target receptor or the biological system that the ligand will ultimately be applied to, mock photo-cross-linking studies in the absence of the photoreactive ligand are carried out. This is an important control as a number of biological chromophores (e.g., tryptophan residues, nucleic acids) and cell types (e.g., retinal cells) are sensitive to various wavelengths of light. Second, to demonstrate that photoinactivation is the direct result of photostimulation of the photoreactive ligand, mock cross-linking studies are conducted with the parent, non-photoreactive ligand. For example, we found that irradiation of AMPARs in the presence of DNQX had no effect on agonist-induced currents. Similarly, photoinactivation studies are carried out with the photoreactive ligand, but using wavelengths of light that do not activate the photoreactive group. In the case of ANQX, we repeated the photoinactivation studies using a neutral density filter to prevent ultraviolet light from reaching the cells and found that under these conditions the antagonist had no lasting effect on agonist-induced currents. Finally, to demonstrate that photoinactivation is selective for the receptor of interest, the effects of the antagonist on other potential receptor targets must be measured. This control is essential if the photoreactive ligand will be used in a biological context (e.g., a neuron), where the activity (or inactivity) of one receptor can influence the activity of other receptors expressed on the cell. In the case of ANQX, we found that the intact ligand also partially inhibited two other glutamate-gated ion channels (i.e., kainate and NMDA receptors), but that photo-cross-linking to these receptors did not occur under the same conditions that resulted in complete inactivation of AMPARs.

3   Summary

Photoinactivation is an effective strategy for rapidly inactivating ligand-gated receptors expressed on the surfaces of cells. The main advantage of this technique over other approaches for inactivating receptors (e.g., gene deletion) is the spatial and temporal resolution it provides. However, the poor penetrance of ultraviolet light in thick tissues necessitates the development of additional photoreactive groups (e.g., two-photon sensitivity) for this technology to be carried forward into such biological preparations.

References

1. Bredt DS, Nicoll RA (2003) AMPA receptor trafficking at excitatory synapses. Neuron 40:361–379

2. Kneussel M, Loebrich S (2007) Trafficking and synaptic anchoring of ionotropic inhibitory neurotransmitter receptors. Biol Cell 99:297–309

3. Lau CG, Zukin RS (2007) NMDA receptor trafficking in synaptic plasticity and neuropsychiatric disorders. Nat Rev Neurosci 8:413–426

4. Michels G, Moss SJ (2007) GABA(A) receptors: properties and trafficking. Crit Rev Biochem Mol Biol 42:3–14

5. Newpher TM, Ehlers MD (2008) Glutamate receptor dynamics in dendritic microdomains. Neuron 58:472–497

6. Shepherd JD, Huganir RL (2007) The cell biology of synaptic plasticity: AMPA receptor trafficking. Annu Rev Cell Dev Biol 23:613–643

7. Adams SR, Tsien RY (1993) Controlling cell chemistry with caged compounds. Annu Rev Physiol 55:755–784

8. Thompson SM, Kao JP, Kramer RH, Poskanzer KE, Silver RA, Digregorio D, Wang SS (2005) Flashy science: controlling neural function with light. J Neurosci 25:10358–10365

9. Ellis-Davies GC (2007) Caged compounds: photorelease technology for control of cellular chemistry and physiology. Nat Methods 4: 619–628

10. Dorman G, Prestwich GD (2000) Using photolabile ligands in drug discovery and development. Trends Biotechnol 18:64–77

11. Chambers JJ, Gouda H, Young DM, Kuntz ID, England PM (2004) Photochemically knocking out glutamate receptors in vivo. J Am Chem Soc 126:13886–13887

12. Sammelson RE, Casida JE (2003) Synthesis of a tritium-labeled, fipronil-based, highly potent, photoaffinity probe for the GABA receptor. J Org Chem 68:8075–8079

13. Chiara DC, Trinidad JC, Wang D, Ziebell MR, Sullivan D, Cohen JB (2003) Identification of amino acids in the nicotinic acetylcholine receptor agonist binding site and ion channel photolabeled by 4-((3-trifluoromethyl)-3H-diazirin-3-yl)benzoylcholine, a novel photoaffinity antagonist. Biochemistry 42:271–283

14. Adesnik H, Nicoll RA, England PM (2005) Photoinactivation of native AMPA receptors reveals their real-time trafficking. Neuron 48:977–985

15. England PM (2006) Rapid photoinactivation of native AMPA receptors on live cells using ANQX. Sci STKE 2006:pl1

16. Cruz LA, Estébanez-Perpiñá E, Pfaff S, Borngraeber S, Bao N, Blethrow J, Fletterick RJ, England PM (2008) 6-Azido-7-nitro-1,4-dihydroquinoxaline-2,3-dione (ANQX) forms an irreversible bond to the active site of the GluR2 AMPA receptor. J Med Chem 51:5856–5860

17. Brauner-Osborne H, Egebjerg J, Nielsen EO, Madsen U, Krogsgaard-Larsen P (2000) Ligands for glutamate receptors: design and therapeutic prospects. J Med Chem 43:2609–2645

18. Bigge CF, Nikam SS (1997) AMPA receptor agonists, antagonists and modulators: their potential for clinical utility. Expert Opin Ther Pat 7:1099–1114

19. Stensbol TB, Madsen U, Krogsgaard-Larsen P (2002) The AMPA receptor binding site: focus on agonists and competitive antagonists. Curr Pharm Des 8:857–872

20. Ohmori J, Sakamoto S, Kubota H, Shimizusasamata M, Okada M, Kawasaki S, Hidaka K, Togami J, Furuya T, Murase K (1994) 6-(1h-Imidazol-1-Yl)-7-nitro-2,3(1h,4h)-quinoxalinedione hydrochloride (Ym90k) and related-compounds—structure-activity-relationships for the AMPA-type non-NMDA receptor. J Med Chem 37:467–475

21. Stein E, Cox JA, Seeburg PH, Verdoorn TA (1992) Complex pharmacological properties of recombinant alpha-amino-3-hydroxy-5-methyl-4-isoxazole propionate receptor sub-types. Mol Pharmacol 42:864–871

22. Jimonet P, Cheve M, Bohme GA, Boireau A, Damour D, Debono MW, Genevois-Borella A, Imperato A, Pratt J, Randle JCR, Ribeill Y, Stutzmann JM, Vuilhorgne M, Mignani S (2000) 8-Methylureido-10-amino-10-methyl-imidazo(1,2-a)indeno(1,2-e) pyrazine-4-ones: highly in vivo potent and selective AMPA receptor antagonists. Bioorg Med Chem 8:2211–2217

23. Kofuji P, Davidson N, Lester HA (1995) Evidence that neuronal G-protein-gated inwardly rectifying K+ channels are activated by G-beta-gamma subunits and function as hetero-multimers. Proc Natl Acad Sci U S A 92:6542–6546

Part III

Chemical Probes for Imaging in the Nervous System

Chapter 10

Development and In Vitro Characterization of Ratiometric and Intensity-Based Fluorescent Ion Sensors

Laurel A. Cooley, Vladimir V. Martin, and Kyle R. Gee

Abstract

Fluorescent ion sensors are quite valuable in experimental biology. The development of new sensor molecules requires determination of spectral properties (absorption bands, fluorescence excitation, and emission maxima) in order to characterize the type of optical response to the target ion. This optical response type and magnitude are used, in combination with solutions of buffered ion of precisely manipulated concentration, to determine the in vitro affinity for the target ion. Buffered aqueous ion solutions of appropriate pH and ionic strength are necessary to predict the performance of new sensors in biological applications. A series of novel benzoxazole-BAPTA calcium sensors, in addition to Rhod-3 (a new version of rhod-2), are described and their optical responses to calcium ion characterized.

Key words Fluorescence, Ion, Calcium, Ratiometric, Sensor, Cells

1 Introduction

It is difficult to overstate the importance of fluorescent ion sensors to experimental biology. In particular, fluorescent sensors for the calcium ion have been critical to neurobiology as optical probes of this important signaling molecule and second messenger, and associated phenomena, in live cells and tissues.

Fluorescent calcium ion sensors were originally developed by Roger Tsien and colleagues in the 1980s (1). The concept involves incorporation of a metal ion chelator within the framework of a fluorescent dye molecule such that ion binding modulates the fluorescent properties of the construct in proportion to the amount of ion present. The calcium ion chelator, known colloquially as BAPTA, is an aromatic analog of ethyleneglycol tetraacetic acid (EGTA). The fluorescence changes are readily measured by fluorescence microscopy, flow cytometry, and fluorescence spectroscopy. The most popular constructs from this era were the ratiometric probes fura-2 and indo-1, (2) and the intensity-based probes

Matthew R. Banghart (ed.), *Chemical Neurobiology: Methods and Protocols*, Methods in Molecular Biology, vol. 995, DOI 10.1007/978-1-62703-345-9_10, © Springer Science+Business Media New York 2013

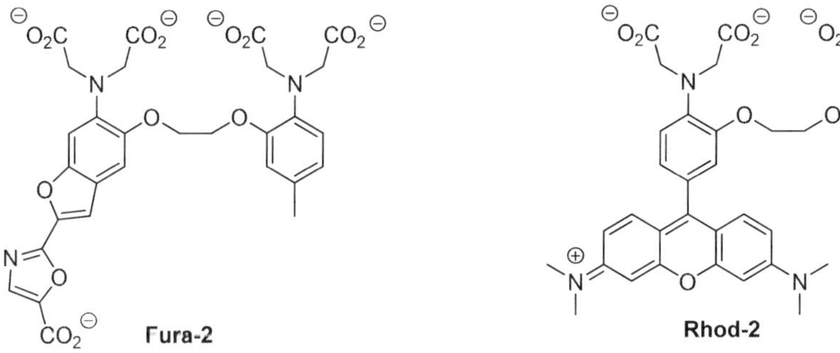

Fig. 1 Chemical structures of *Fura-2* and *Rhod-2*

fluo-3 and rhod-2 (Fig. 1) (3). The ratiometric versions experience a fluorescence excitation (fura-2) or emission (indo-1) wavelength change upon ion binding. The intensity-based versions report a fluorescence emission intensity change (typically increase) upon ion binding but without a significant wavelength change. The virtues and relative merits of these types of compounds have been extensively reviewed (4).

The value of these fluorescent ion sensor tools inspired the development of new versions with improved properties (5), and features tailored to more specific biological applications (6, 7). It may seem obvious, but development of new fluorescent ion sensors typically involves design and synthesis of new molecules that are envisioned to have certain optical properties. After synthesis (which is typically multistep) and purification, the first key characterization step is to evaluate the fluorescence properties in vitro. This first involves determination of the absorption spectra in aqueous buffer. Next the fluorescent properties are determined in the absence and presence of ion at concentrations expected to saturate the sensor at equilibrium. At this point a qualitative determination can be made as to whether the fluorescence of the new sensor molecule is responsive to the target ion, and if so, whether that response is in the form of excitation ratiometry, emission ratiometry, or emission intensity change. Sensors that exhibit excitation or emission ratiometry upon ion binding can be calibrated using a ratio of the fluorescence intensities at two different wavelengths, resulting in the cancellation of artifactual variations in the fluorescence signals that might otherwise be misinterpreted as changes in ion concentration. Genuine fluorescence emission ratiometry is poorly understood and quite rare, but highly prized because of this "internal standard" property that can be achieved with a single excitation source.

The second key in vitro characterization step is determination of the ion binding affinity, typically expressed as the dissociation constant (K_d). The concentration range over which a sensor produces an observable response is approximately $0.1 \times K_d$ to $10 \times K_d$. Thus in the sensor molecule design, an ion affinity is targeted based on the expected biology application. The first generation of probes (2, 3) was well suited to monitor changes in cytosolic $[Ca^{2+}]$ from a resting state to stimulated. For use in situations in which these sensors are saturated with relatively high $[Ca^{2+}]$, such as those elicited by action potentials, sensor affinity can be attenuated. For fluorescent calcium ion sensors, this is often accomplished by strategic attachment of electron-withdrawing substituent(s) to the BAPTA phenyl ring opposite the phenyl ring to which the fluorophore is attached (8, 9); the more potent the electron-withdrawing substituent, the lower the ion binding affinity (and the higher the K_d value). After synthesis, K_d calibration procedures consist of recording fluorescence signals from the new probe, corresponding to a series of precisely manipulated ion concentrations. The resulting sigmoidal titration curve is either linearized by means of a Hill plot or analyzed directly by nonlinear regression to yield K_d values. For in vitro characterization of calcium ion sensors, EGTA buffering is necessary to produce defined calcium ion concentrations that can be calculated from the K_d of the Ca^{2+}–EGTA complex (10).

To exemplify these in vitro spectroscopic characterization techniques, described herein are the spectral and K_d characterization experiments for a small series of novel fluorescent calcium ion sensors: excitation ratiometric (**compound 1**), emission ratiometric (**compound 2**), and fluorescence intensity-based (**compounds 3, 4**) versions (Fig. 2). Compounds **1–3** are members of a novel fused benzoxazole-BAPTA family in which the spectroscopic properties vary dramatically depending on the substituent attached between the nitrogen and oxygen atoms of the benzoxazole. Sensor **4**, known as Rhod-3, represents an attempt to improve upon the live cell cytosolic localization properties of Rhod-2.

2 Materials

(a) Calcium calibration buffers (Life Technologies) which contain 10 mM K_2EGTA and 10 mM CaEGTA. Store at 4 °C for up to 3 months. Both of these solutions contain 100 mM KCl and 30 mM MOPS, pH 7.2. These will be blended (described in next section) to make buffers which will have a free Ca^{2+} range from 0 to 39.8 μM.

(b) Appropriate calcium indicator. The $[Ca^{2+}]_{free}$ in the 10 mM CaEGTA is high enough to saturate indicators with K_d values in the 100–1,000 nM range. These indicators include fura-2, indo-1, fluo-4, and rhod-2. Since the new sensors 1–4 are based on the same BAPTA framework, without electron-withdrawing substituents, it is anticipated that this calcium ion

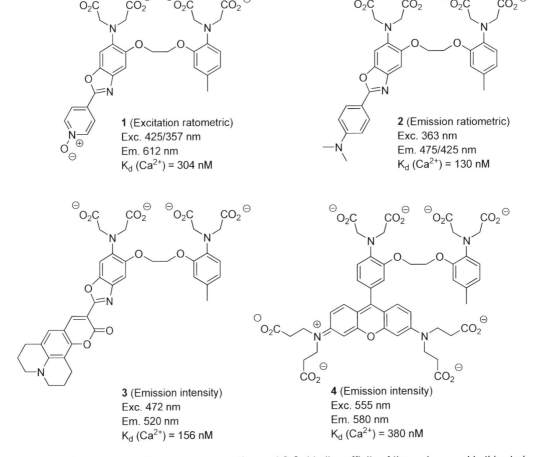

Fig. 2 Chemical structures, fluorescence properties, and Ca^{2+}-binding affinity of the probes used in this study

concentration range will be adequate to properly evaluate the new sensors' K_d values.

(c) Methyl acrylate disposable cuvettes (VWR).

(d) UV-Vis quartz spectrometry cells (matched pair) (Starna Cells, Inc.)

3   Methods

(a) Prepare an approximately 1 mM stock solution of the fluorescent sensor in any Ca^{2+}- and EGTA-free solvent such as e-pure water, dimethyl sulfoxide, methanol, or acetonitrile. Avoid solvents which are not miscible in water, such as chloroform.

(b) Prepare a working solution of the Ca^{2+} indicator from the 1 mM stock solution prepared in Subheading 3, step (a), with

Table 1
Preparation of buffered Ca²⁺ solutions

Free Ca²⁺ (µM)	mL of 10 mM K₂EGTA ("zero Ca²⁺ sample")	mL of 10 mM CaEGTA ("high Ca²⁺ sample")
0.0	5.0	0.0
0.017	4.5	0.5
0.038	4.0	1.0
0.065	3.5	1.5
0.100	3.0	2.0
0.150	2.5	2.5
0.225	2.0	3.0
0.351	1.5	3.5
0.602	1.0	4.0
1.35	0.5	4.5
39.8	0.0	5.0

the zero calcium buffer (10 mM K₂EGTA), generally about 60 µM.

(c) If the excitation wavelength is unknown, generate an absorption spectrum using an adequate amount of the working solution for the UV-Vis quartz spectrometry cells. The absorption maximum of this spectrum will be used as the fluorescence excitation parameter.

(d) Prepare a series of free Ca²⁺ solutions according to Table 1 (see Notes 1 and 2).

A precise set of fluorescence curves can be obtained for determining the dissociation constant (K_d) by varying the free Ca²⁺. It is important to keep the indicator concentration, pH, ionic strength, and temperature constant (see Notes 3 and 4).

(e) Instrument and experiment parameters will need to be optimized for the sample. Prepare a trial dilution of the working solution into 5 mL of the 10 mM CaEGTA ("high Ca²⁺ buffer"). To determine the emission maximum and appropriate spectral range, excite the sample at the wavelength determined by your sensor's absorption maximum. Alter the instrument parameters to obtain the best range, slit wide, and speed. Increase or decrease the sample concentration to obtain the best spectra knowing that this will be the largest peak of interest. If the emission band intensity is either too high or too low then some alterations to the sample must occur. Try changing

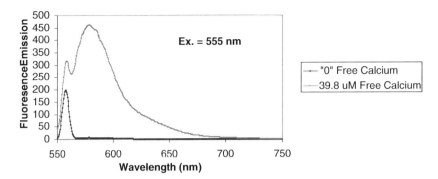

Fig. 3 Fluorescence spectra of sensor **4** in the absence and presence of "high," i.e., saturating, calcium ion (excitation at 555 nm)

the slit widths of the instrument or change the concentration of the sample until an acceptable emission intensity is obtained.

(f) Prepare a "0" free calcium sample at the concentration determined in step (e) and obtain the emission spectrum using the same parameters as used in step (e).

(g) Compare the spectra obtained for the "high-calcium buffer" (39.8 μM free calcium) and zero free calcium. For the sample to be a calcium sensor, there must be a difference in the fluorescence intensity and/or a shift in the wavelength (for a ratiometric indicator). Notice the significant difference in signal in Fig. 3 between the zero free calcium and the 39.8 μM free calcium samples at the same concentration of the calcium sensor **4**. The high-calcium sample has a fluorescence emission intensity of approximately 450 relative fluorescence units (rfu). For the particular instrument used for this experiment the highest intensity is 1,000. Any peak intensity over 1,000 will be cut off above that point, so 500 is a good mid-range target for this instrument (see Note 5). Another consideration is the effect of light scattering at the excitation wavelength. In this case, the emission was excited at 555 nm and the peak seen at 555 nm is the result of light scattering on this instrument and is not an emission maximum of the calcium sensor. The emission maximum for the spectra in Fig. 3 is at approximately 580 nm.

To determine a preliminary K_d value, a stock solution of 17 μM **4** (as the tri-potassium salt) in methanol was prepared. From the results of the spectra obtained using the "high"-calcium buffer and the zero calcium buffer (Fig. 3) it was determined that an 84.8 nM solution in each buffer solution was the optimum concentration to collect the series of spectra associated with Fig. 4. (This procedure is described in more detail in Subheading 3, step (h).) (see Note 6).

Dilutions were made using 25 μL of the stock solution into 5 mL of each of the series of free calcium buffers prepared

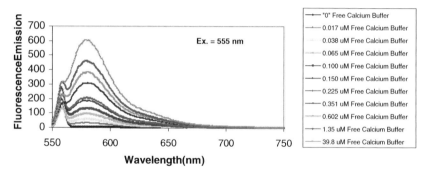

Fig. 4 Fluorescence emission spectra of sensor **4** in buffer with calcium ion concentration of 0–39.8 μM (excitation at 555 nm)

Table 2
Fluorescence titration data for sensor 4 with increasing [Ca²⁺]

$[Ca^{2+}]_{free}$ (μM)	F (580.4 nm)	$\log([Ca^{2+}]_{free})$	$\log\{(F-F_{min})/(F_{max}-F)\}$
0.0	5.61		
0.017	30.8	−1.77	−1.354
0.038	59.59	−1.42	−1.001
0.065	96.44	−1.187	−0.744
0.1	134.52	−1	−0.558
0.15	181.67	−0.824	−0.376
0.225	204.43	−0.648	−0.299
0.351	306.57	−0.455	0.011
0.602	383.28	−0.22	0.241
1.35	457.22	0.13	0.5
39.8	600.05		

in step (d). The sample was excited at 555 nm and the emission maximum was determined to be 580.4 nm. The intensity of the spectra for each of the free calcium buffers was recorded at the emission maximum in order to determine the K_d of the calcium indicator. See Table 2.

The $\log([Ca^{2+}]_{free})$ was plotted against $\log\{(F-F_{min})/(F_{max}-F)\}$ using Microsoft Excel in Fig. 5.

The K_d for the calcium indicator **4** can be determined from the equation for the line. Set $y=0$ and solve for x. x is the log of the K_d value. The data for this calcium indicator result in a measured K_d of 380 nM.

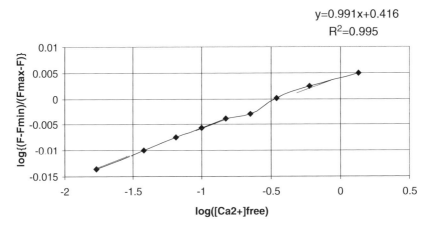

Fig. 5 Determination of K_d for calcium indicator **4** by plotting $\log([Ca^{2+}]_{free})$ vs. $\log\{(F-F_{min})/(F_{max}-F)\}$

(h) The K_d of the indicator can be best calculated once the optimum instrument parameters have been determined for range, slit width, speed, and sample concentration. For example, the emission spectra in Fig. 4 for sensor **3** were obtained on a Perkin Elmer LS 55 Luminescence Spectrometer. The range was 485–700 nm; SW = 5 nm/5 nm; and speed = 300 nm/min. The 51 nM sample in the "high-Ca²⁺ buffer" was excited at 472 nm as determined by its absorption maxima. Notice that the range was optimized to run from 485 to 700 nm. This cut off the light scattering peak from the excitation at 472 nm and collected spectral information until it reached the baseline at 700 nm.

(i) To collect the spectra, prepare each free calcium sample including a fresh sample of both the "high-calcium buffer" (39.8 μM free calcium) and the zero calcium buffer. The fresh samples are necessary in the event that the fluorescent sensor undergoes photobleaching during analysis. If careful dilutions have been made and the sensor undergoes an emission shift as shown in Fig. 6, the result will be a clean isosbestic (crossover) point.

(j) If an excitation ratiometric K_d is desired, optimum parameter and sample concentration are determined similarly after determining the emission maximum. See Fig. 7 for an example of ratiometric excitation spectra.

(k) Once the excitation or emission spectra have been collected, record the values of the spectral maximum for each free calcium sample. These will be plotted against the $[Ca^{2+}]_{free}$ to obtain a calibration curve which is used to determine the dissociation constant. See Table 3 below which uses the data collected from Fig. 8.

(l) The data are plotted with the $\log[Ca^{2+}]_{free}$ on the x-axis and the $\log\{(F-F_{min})/(F_{max}-F)\}$ on the y-axis which yields the calibration curve. See Fig. 6.

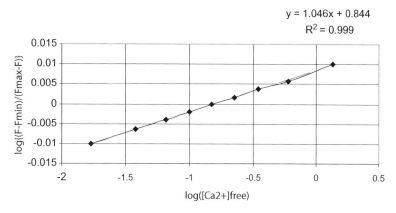

Fig. 6 Determination of K_d for calcium indicator *3* by plotting $\log([Ca^{2+}]_{free})$ vs. $\log\{(F-F_{min})/(F_{max}-F)\}$

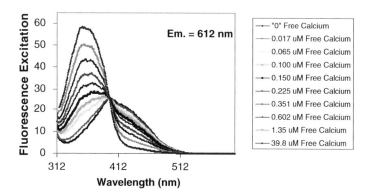

Fig. 7 Fluorescence excitation spectra of sensor **1** in buffer with calcium ion concentration of 0–39.8 µM (emission collected at 612 nm)

(m) The above data were graphed using Microsoft Excel which includes the equation for the line. As a linear "double-log" plot, the *x*-intercept is equal to the log of the K_d for the calcium indicator. To solve, set $y=0$. The result is $x=-0.8069$. Since *x* is the log of the K_d indicator, the result is $10^{-0.8069}$, or antilog, which yields a K_d value of 156 nM.

　　The following Web site has a tool for determining the K_d of your data: http://www.invitrogen.com/site/us/en/home/support/Research-Tools/KD-calculator.html.

(n) Disassociation constants can also be determined from ratiometric indicator emissions or excitation data. See Fig. 9. The ratio of the two wavelengths can be plotted against $[Ca^{2+}]_{free}$.

(o) Table 4 shows the data for the emission ratiometric sensor **2**.

(p) The data are plotted with the $\log[Ca^{2+}]_{free}$ on the *x*-axis and the $\log\{((R-R_{min})/(R_{max}-R))(F_{2min}/F_{2max})\}$ on the *y*-axis which yields the calibration curve. See Fig. 10.

Table 3
Fluorescence titration data for sensor 3 with increasing [Ca²⁺]

$[Ca^{2+}]_{free}$ (µM)	F (520.10 nm)	$\log([Ca^{2+}]_{free})$	$\log\{(F-F_{min})/(F_{max}-F)\}$
0.0	4.02		
0.017	44.42	−1.77	−1.005
0.038	89.04	−1.42	−0.631
0.065	131.31	−1.187	−0.402
0.100	177.18	−1	−0.202
0.150	224.5	−0.824	−0.015
0.225	271.15	−0.648	0.167
0.351	318.01	−0.455	0.367
0.602	358.48	−0.22	0.575
1.35	413.03	0.13	1.012
39.8	452.86		

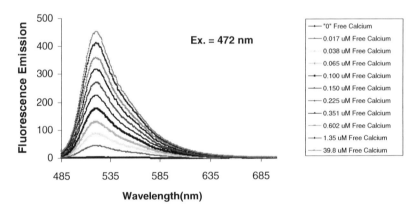

Fig. 8 Fluorescence emission spectra of sensor **3** in buffer with calcium ion concentration of 0–39.8 µM (excitation 472 nm)

(q) Using the equation for the line as in Subheading 3, step (1), the K_d of the emission ratiometric sensor **2** is 130 nM. Although the calculations of the K_d for ratiometric sensors are more complicated, these measurements reduce factors that can influence the K_d calculation. These factors include slight differences in the sensor concentration, photobleaching, and any pathlength differences in the disposable cuvettes. Because these differences affect both intensities similarly at both wavelengths, the ratio of the two intensities effectively cancels those factors out.

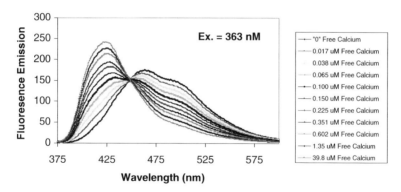

Fig. 9 Fluorescence emission spectra of sensor **2** in buffer with calcium ion concentrations of 0–39.8 μM (excitation 472 nm)

Table 4
Fluorescence ratiometry of sensor 2 with increasing [Ca²⁺]

$[Ca^{2+}]_{free}$ (μM)	F_1 (425.06 nm)	F_2 (463.92 nm)	R	$\log([Ca^{2+}]_{free})$	$\log\{((R-R_{min})/(R_{max}-R))(F_{2min}/F_{2max})\}$
0.0	74.09	175.11	0.423		
0.017	92.01	165.28	0.557	−1.77	−0.899
0.038	111.55	158.66	0.703	−1.42	−0.542
0.065	130.65	150.57	0.868	−1.19	−0.298
0.100	148.04	142.47	1.039	−1.00	−0.105
0.150	163.26	134.08	1.218	−0.82	0.065
0.225	181.11	126.12	1.436	−0.65	0.257
0.351	193.31	118.11	1.637	−0.45	0.434
0.602	213.4	114.33	1.867	−0.22	0.66
1.35	227.49	106.92	2.128	0.13	1.006
39.8	240.7	99.3	2.424		

4   Notes

1. The table for preparing the series of $[Ca^{2+}]_{free}$ solutions in Subheading 3, step (d), makes 5 mL of each solution. Before starting the dilutions, be sure to have at least 28 mL of the 10 mM K₂EGTA and 33 mL of the 10 mM CaEGTA. The disposable cuvettes listed in Subheading 2 have volume of 4.5 mL. If another size of cuvette is used, the quantities of the 10 mM K₂EGTA and 10 mM CaEGTA can be scaled proportionately.

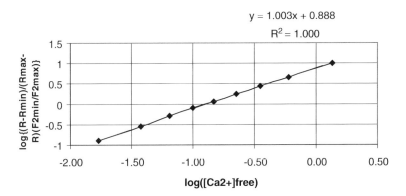

Fig. 10 Determination of K_d for calcium indicator **2** by plotting $\log([Ca^{2+}]_{free})$ vs. $\log\{((R-R_{min})/(R_{max}-R))$ $(F_{2min}/F_{2max})\}$

2. Allow a minimum of 1 h to prepare each dilution series (of 11 solutions) with increasing $[Ca^{2+}]_{free}$ and to produce the corresponding spectra. This is the time needed after all the instrument and concentration parameters for the experiment have been determined.

3. The e-pure water used is deionized water (resistance 18 MΩ) and does not contain any KCl or MOPS.

4. It is sometimes useful to perform the described K_d determination on a known fluorescent ion sensor first and compare the results with known values prior to analyzing a novel sensor molecule.

5. The $[Ca^{2+}]_{free}$ is calculated from the K_d of EGTA for Ca^{2+}:

$$\left[Ca^{2+}\right]_{free} = K_d^{EGTA} \times \left[CaEGTA / K_2EGTA\right].$$

When 4.5 mL of the K_2EGTA is blended with 0.5 mL of CaEGTA (see Table 1 at Subheading 3, step (d)), the ratio of CaEGTA to K_2EGTA is 0.5 mL:4.5 mL or 1:9 which equals 0.11. The calcium buffer in the kit (Life Technologies) has a pH of 7.2, ionic strength of 100 mM KCl. At 20 °C, the K_d^{EGTA} is 150.5×10^{-9} M (11). Therefore

$$\left[Ca^{2+}\right]_{free} = \left(150.5 \times 10^{-9}\,M\right) \times 0.11$$
$$= 1.67 \times 10^{-8}\,M$$
$$= 0.0167\,\mu M$$
$$= 0.017\,\mu M.$$

6. For ion sensors, the K_d value is quite sensitive to a number of environmental factors, including temperature, pH, and ionic strength. The conditions described herein for K_d determination are chosen to approximate those of a cell interior. Most

often for neurobiology experiments, these sensor molecules are intended for use after loading into live cells. Interaction with intracellular proteins and other components is difficult to model with an in vitro system, even though such interactions can also affect K_d values (usually upward). Thus it is important to determine sensor K_d values in situ within cells if quantitative ion concentration values are desired in live cell systems.

Acknowledgment

The authors would like to thank Dr. Iain Johnson for helpful advice and guidance.

References

1. Tsien RY (1980) New calcium indicators and buffers with high selectivity against magnesium and protons: design, synthesis, and properties of prototype structures. Biochemistry 19: 2396–2404

2. Grynkiewicz G, Poenie M, Tsien RY (1986) A new generation of Ca^{2+} indicators with greatly improved fluorescence properties. J Biol Chem 260:3440–3450

3. Minta A, Kao JP, Tsien RY (1989) Fluorescent indicators for cytosolic calcium based on rhodamine and fluorescein chromophores. J Biol Chem 264:8171–8178

4. Haugland RP (2005) The handbook. A guide to fluorescent probes and labeling technologies, 10th edn. Invitrogen Corp., San Diego, CA, Chapter 19 and references cited therein

5. Gee KR, Brown KA, Chen WN-U, Bishop-Stewart J, Gray D, Johnson I (2000) Chemical and physiological characterization of Fluo-4 Ca^{2+}-indicator dyes. Cell Calcium 27:97–106

6. Beierlein M, Gee KR, Martin VV, Regehr W (2004) Presynaptic calcium measurements at physiological temperatures using a new class of dextran-conjugated indicators. J Neurophysiol 92:591–599

7. Gee KR, Zhou Z-L, Qian W-J, Kennedy R (2002) Detection and imaging of zinc secretion from pancreatic β-cells using a new fluorescent zinc indicator. J Am Chem Soc 124:776–778

8. Hollingworth S, Gee KR, Baylor SM (2009) Low affinity Ca^{2+} indicators compared in measurements of skeletal muscle Ca^{2+} transients. Biophys J 97:1864–1872

9. Gee KR, Archer EA, Lapham LA, Leonard ME, Zhou Z-L, Bingham J, Diwu Z (2000) New ratiometric fluorescent calcium indicators with moderately attenuated binding affinities. Bioorg Med Chem Lett 10:1515–1518

10. Kao JPY (1994) Practical aspects of measuring [Ca^{2+}] with fluorescent indicators. Methods Cell Biol 40:155–181

11. Tsien RY, Pozzan T (1989) measurement of cytosolic free Ca^{2+} with Quin2. Methods Enzymol 172:230–262

Chapter 11

Characterization of Voltage-Sensitive Dyes in Living Cells Using Two-Photon Excitation

Corey D. Acker and Leslie M. Loew

Abstract

In this protocol, we describe the procedures we have developed to optimize the performance of voltage-sensitive dyes for recording changes in neuronal electrical activity. We emphasize our experience in finding the best dye conditions for recording backpropagating action potentials from individual dendritic spines in a neuron within a brain slice. We fully describe procedures for loading the dye through a patch pipette and for finding excitation and emission wavelengths for the best sensitivity of the fluorescence signal to membrane voltage. Many of these approaches can be adapted to in vivo preparations and to experiments on mapping brain activity via optical recording.

Key words Voltage-sensitive dye, Dendritic spine, Membrane potential, Two-photon imaging, Photostability

1 Introduction

Optical recording techniques provide powerful tools for neurobiologists to study signaling over time and space in living cells and tissues (1). Voltage-sensitive dye (VSD) imaging permits rapid access to membrane potential signals corresponding to action potential propagation and integration of synaptic activity in neuronal dendrites. VSDs have been shown to be capable of recording very brief, *superthreshold* membrane potential "spikes" (~1 ms) from very small structures such as dendritic spines (~1 μm in diameter, (2–4)) which would be inaccessible by patch clamp recording (Fig. 1). Limitations still exist to achieving what we consider the ultimate goal of VSD recordings: high-fidelity, single-sweep recordings of *subthreshold* synaptic potentials in single spines. Consequently, VSD development is ongoing to improve the sensitivity and signal-to-noise quality of these recordings. As with any optical probe, details of the biophysical function and interaction of the molecule determine its ability to act as a signal reporter. Here

Matthew R. Banghart (ed.), *Chemical Neurobiology: Methods and Protocols*, Methods in Molecular Biology, vol. 995, DOI 10.1007/978-1-62703-345-9_11, © Springer Science+Business Media New York 2013

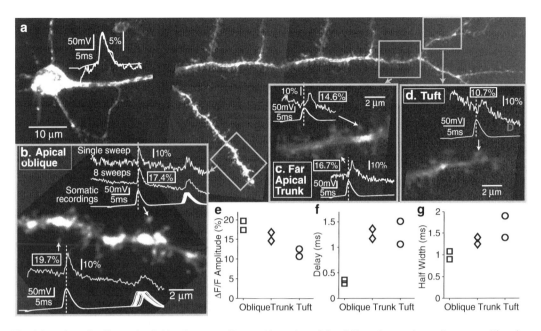

Fig. 1 Imaging of action potential backpropagation and invasion of dendritic spines using voltage-sensitive dye (VSD) di-2-AN(F)EPPTEA. (**a**) An image montage of a cortical pyramidal neuron loaded via somatic patch pipette with the VSD. The traces at the *upper left* show superimposed somatic electrical and perisomatic VSD records, both at 10 kHz, demonstrating precise temporal synchronization. (**b**) Recordings of backpropagating action potentials (bAPs, two spikes, analysis applies to first spike only), elicited using somatic current injection, at two different spines in the same region on an apical oblique dendrite. Somatic electrical waveforms with multiple recordings aligned at the first spike are shown below single-voxel imaging data from spines. *Inset boxes* show the amplitudes of the optically recorded bAP waveforms. For the *top* spine, a single optical sweep is shown along with the average of eight temporally aligned sweeps. Single sweep measurements: 18.0% amplitude, 0.29-ms delay, 0.32-ms rise time, 0.83 ms half-width, SNR = 8.3. (**c**) Same as panel (**b**), except that two spines are targeted on a distal apical trunk region. (**d**) VSD recording from a spine on the apical tuft. (**e–g**) Summary of amplitudes, propagation delay times (optical relative to electrical peak time), and half-widths for bAPs recorded in spines in all three regions. In all cases, spines (indicated by *arrows*) were placed in focus and targeted at their centers. *Vertical dashed lines* indicate peak times of somatic action potentials. Laser power (measured after objective, above slice): 4–5.5 mW (proximal to distal locations). Reproduced from (3), with permission from Elsevier

we describe our strategy to assess the "performance" of VSDs by recording backpropagating action potential (bAP) waveforms in single dendritic spines at 10 kHz using two-photon excitation microscopy.

Many different aspects of VSDs can affect their ability to act as a voltage reporter (5). Conveniently, most VSDs exhibit some important properties that are sometimes taken for granted, but which should be considered for every newly developed dye. For example, most VSDs linearly report membrane potential over the physiological range of approximately –70 to 50 mV. Similarly, most VSDs exhibit very fast kinetics when recording fluorescence, accurately reflecting the action potential waveform. Further, they do not produce pharmacological or toxic effects over the range of concentrations required for good signal-to-noise. Still, for any new dye or new experimental preparation, tests and controls must be performed to assure that these conditions are met.

Newly emerging genetically encoded, potential sensing probes (6–8) carry inherent advantages, since they can be targeted to specific neuronal populations, but there are important trade-offs to consider. For example, their poor sensitivity and slow kinetics do not permit these probes to accurately follow brief membrane potential fluctuations such as action potentials. However, new proteorhodopsin-based probes show improved voltage sensitivity and speed capabilities (9), and show great promise, especially if these attributes can be preserved while improving their fluorescence quantum efficiencies.

Practically speaking, a good VSD must provide sufficient sensitivity so that recordings of signals of interest are obtained with sufficient signal-to-noise (S:N) ratio for the demands of the experiment. This means that the signal (i.e., the voltage-dependent change in the optical signal) component of the recording should be large compared to inherent noise in the recording. In optical recordings, good S:N is achieved first by generating a large number of photons from the optical probe and second by having a large sensitivity whereby even a small change in the membrane potential results in a significant change in the number of photons. Since the number of photons produced depends on the amount of indicator, VSD concentration and therefore loading procedures are critical. Other factors are also important, such as the fraction of the VSD that binds to the plasma membrane vs. non-excitable cell or internal membranes vs. the dye concentration that stays in solution; the latter populations of dye will produce background fluorescence that will reduce the overall sensitivity of the dye to voltage changes on the plasma membrane. However, for the VSDs developed in our lab, fluorescence from aqueous dye molecules is insignificant because they have very low fluorescence quantum efficiency in water, so only dye staining of organelle membranes will be of concern. With a given concentration of dye, more photons can be generated by increasing the excitation intensity to yield a better S:N. However, since every excited dye molecule has a probability of undergoing irreversible photobleaching, the overall fluorescence will decay excessively when excitation is too large. VSD photostability depends on the properties of the dye and is an active area of research to optimize dye performance; this is because dyes that are less likely to undergo photobleaching can produce more fluorescent photons to yield better S:N at a given dye concentration.

Finally, it is also important to emphasize that the dye response can be critically sensitive to the choice of excitation and emission wavelength. The dyes developed in our lab generally undergo a red shift in their spectra in response to membrane depolarization when they are bound to the cytoplasmic side of the plasma membrane (i.e., internal staining); a blue-shift is produced when the dyes are applied to the cell exterior. Therefore the optimal wavelength for the maximal response to voltage is at the edges of the spectral band—somewhat counterintuitively, the dyes show the smallest sensitivity to voltage near the peak of the excitation and emission spectra. Of course, if excitation and emission are too far to the edge of the spectra the overall fluorescence intensity will be too low to allow for a good signal-to-noise for the measurement. Thus the excitation and emission wavelengths should be carefully optimized for a given dye and a given experimental setup.

Here we describe how VSDs can be loaded into single cells in brain tissue through the patch pipette in a controlled manner (10) for a more reproducible final concentration within the cell. We then describe our method of recording bAPs in spines and how we can quantify these data to determine the dye's voltage sensitivity, the attainable signal-to-noise, and photobleaching characteristics. Loading via the patch pipette can be variable since it depends on the access resistance of the whole-cell patch. Alternative methods such as single-cell electroporation (11) have not yet been tested for reproducibility with VSDs. Additionally, dyes that have reduced solubility may diffuse slower and produce a nonuniform intracellular distribution, where the local concentration at any particular region of interest may be significantly different than at another, for example more distal, region. These limitations in our ability to reproducibly control dye loading limit our ability to quantitatively assess dye performance or to demonstrate clear advantages of some dyes over others that may only differ slightly on the molecular level. In spite of these challenges, we find that the following approach is a useful way to assess the suitability of VSDs and importantly uses a consistent biological signal in live tissue as a test bed.

2 Materials

1. Millex Nonsterile Syringe Filters, 0.22 μm pore (Millipore).

2. Microcentrifuge tubes, 0.6 ml (VWR).

3. Small centrifuge (VWR).

4. Original Eppendorf microloaders, 20 μl, connected to a 3 ml syringe by a female luer, 1/6 in. O.D. fitting (World Precision Instruments).

5. Osmometer (Westcor).

6. Sonicator (VWR).

7. Manometer (SPER Scientific).

8. Borosilicate glass capillaries, Sutter 1.0 O.D. thin wall (0.78 I.D.) with filament (Sutter).

9. Pipette puller (Sutter).

10. 1 mM (or greater) stock dye solution in ethanol. Examples: JPW3028 (di-1-ANEPEQ), JPW1114 (di-2-ANEPEQ), PY3243 (di-2-AN(F)EPPTEA).

11. Custom 2-photon microscope equipped for electrophysiology. Our system consists of a Zeiss AxioSkop 2FS MOT microscope (Zeiss) adapted to a Chameleon Ultra II Ti:Sapphire laser (Coherent), fit with a 40× 1.0 NA water immersion objective (Zeiss) and equipped for IR-DIC imaging. Electrophysiological recordings are obtained using an Axopatch 1D electrophysiology amplifier (Axon Instruments).

12. Electro-optic modulator (EOM) with BK (resonance-damp-ened) option (Conoptics).

13. ScanImage MATLAB-based laser scanning software (12), MATLAB based.

14. Additional ScanImage-compatible software for recording single spots or "voxels" such as VoxelRecordVSD.m (3), MATLAB based, compatible with ScanImage.

15. Computer controlled current-sensitive preamplifier (Stanford Research).

16. Acute rodent brain slices for electrophysiological analysis (13).

17. Artificial cerebrospinal fluid (ACSF) containing 127 mM NaCl, 25 mM $NaHCO_3$, 25 mM d-glucose, 2.5 mM KCl, 1.25 mM NaH_2PO_4, 2 mM $CaCl_2$, and 1 mM $MgCl_2$ (pH 7.3 with carbogenation).

18. Intracellular solution containing 135 mM K-gluconate, 2 mM $MgCl_2$, 2 mM Mg-ATP, 10 mM Na-phosphocreatine, 0.3 mM Na-GTP, 10 mM HEPES, and 0.01 mM EGTA (pH 7.4 with KOH).

3 Methods

3.1 Preparation and Controlled Loading of Voltage-Sensitive Dyes

1. Prepare acute brain slices and place under microscope objective while perfusing with ACSF at room temperature.

2. Prepare VSD dissolved in intracellular solution from ethanol stock solution. First dry sufficient dye from the stock solution to pure solid by gently blowing in inert gas such as argon (may take >15 min). Add intracellular solution to make 500 μM–1 mM final solution, sonicate, and filter using a microcentrifuge and syringe filter. Check that intracellular osmolarity is ~30 mosmol less than extracellular.

3. Load a minimum amount of dye-free intracellular solution into the tip of patch pipette, significantly less than the pipette tip volume, ~0.5–1 μl. Any extra that can be removed with suction

should be removed. Load the dye solution with a separate microloader. Preloading with dye-free solution minimizes the spillage of dye into the slice prior to patching, which would not wash out.

4. Patch a neuron (14) taking care not to spill any dye. Use minimal, ~ 6 mbar, positive pressure until entering the slice, only increasing pressure when ready to penetrate the slice and approach the cell, ~30 mbar (see Note 1).

5. After successfully patching, image in 2-photon mode at the peak 2-photon absorption wavelength of the dye (920 nm for di-2-AN(F)EPPTEA). Initial images should show fluorescence at the pipette tip, but no staining of the surrounding tissue (which would be indicative of spillage) and very little or no staining of the cell.

6. Monitor the increase in staining of cell body in 2-photon until a criterion is met in terms of image intensity. Here it is crucial to control all aspects of the scanning such as zoom, scan speed, sample rate, etc., along with laser wavelength and intensity. (see Note 2) When the final intensity is achieved, the patch is removed by slowly pulling off and forming an outside-out patch, which ensures that the cellular membrane reseals.

7. After waiting for a sufficient time to allow for adequate diffusion of the dye (30 min is often enough for JPW3028 or PY3243), repatch the neuron with a new pipette loaded with normal intracellular solution and maintain the neuron at a negative membrane potential in current clamp mode to prevent inadvertent action potential firing ((10), see Notes 3–6). Cell appearance should be virtually identical before and after filling under IR-DIC (Fig. 2).

3.2 Spine-Targeted, Single-Voxel Recordings at 10 kHz

1. Grab a frame with a region of interest (roi) including spines using ScanImage.

2. With spine of interest in focus, select the target by selecting the desired pixel from the frame scan. Spines can be targeted at their center, which seems to maximize reproducibility, although voltage sensitivity might be slightly higher if a more peripheral spot is selected. Convert the pixel location to the corresponding laser position or voltage command for the galvos. Targeting and recording are performed using VoxelRecordVSD.m, a MATLAB graphical user interface written in our lab, which works in conjunction with ScanImage (screenshot, Fig. 3). Target position is saved along with fluorescence data.

3. Using an external trigger generated by the electrophysiology system that corresponds to the injection of ~200–300 pA of current into the neuron to generate a single action potential (AP) or doublet, move the laser position to the spine, open the EOM, and record the detector output for a period of time, usually 15–20 ms pre-AP, and 40–50 ms total for single APs or doublets.

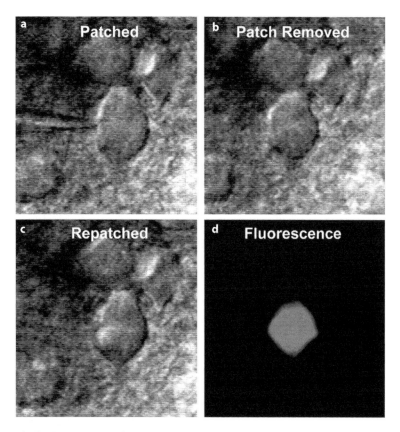

Fig. 2 Sequence for loading neurons with voltage-sensitive dyes (VSDs) via whole-cell patch pipettes. (**a**) Initial patch pipette can be seen to the *left* of the neuron in the center of the frame visualized with IR-DIC optics. (**b**) After a target level of somatic staining is reached, the patch electrode is carefully removed. The neuron should retain its original appearance. (**c**) Neuron is then repatched after a delay to allow for diffusion of the dye into dendrites and spines. (**d**) Bright fluorescence to demonstrate that the cell is loaded with VSD

4. Filtering: Before sampling at 10 kHz, the signals must be low-pass filtered with a cutoff frequency below 5 kHz. This is important for signal-to-noise in optical recordings since raw signals from PMTs are very-high-bandwidth, noisy signals. We use a computer-controlled preamp to change the cutoff frequency to the closest setting of 3 kHz. Typically, a filter setting of 300 kHz, which passes much more noise, is necessary for frame or line scans where pixel times are ~3 µs (e.g., 2 ms/line at 512 pixels/line). In an automated system, the filter setting is returned to the original, higher setting after recording from voxels and before more frames are acquired.

5. Laser power: For VSD recordings, the laser power required for adequate frame scans of spines and dendrites was also appropriate for single-voxel recordings. Measured at the surface of the slice, powers for spine recordings were typically 3–5 mW.

6. Background can be measured by choosing targets and recording from voxels to the side of or away from targeted spines using

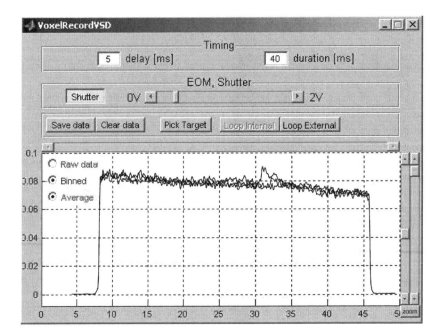

Fig. 3 Screen shot of MATLAB graphical user interface for single-voxel, optical recordings of membrane potential. Imaging duration and delay following the trigger signal from the electrophysiology system can be controlled, along with the electro-optic modulator (EOM), which serves as a shutter and a means to control the power

the same laser intensity. In our experiment, the background originates primarily from the electronics, usually less from detection of excitation light, but not at all from background fluorescence as long as care was taken not to spill any dye (see above).

7. Although the detector output could be directly sampled at 10 kHz at this point, we typically oversample at 1 MHz and then downsample or "decimate" back to 10 kHz in an effort to further improve the signal-to-noise.

3.3 Characterizing Voltage Sensitivity, Signal-to-Noise, and Photobleaching of Voltage-Sensitive Dyes by Recording Backpropagating Action Potentials in Single Dendritic Spines

1. *Raw data collection using an "alternating" protocol with interspersed control trials (refer to Fig. 4).* Record trials with single bAPs as described above along with control trials in order to measure bleaching and signal-to-noise. Control trials and action potentials should ideally be alternated or intermixed. Temporal jitter can be seen between trials when eliciting action potentials, which must be accounted for using "spike-triggered averaging" (step 2, below). Bleaching rate depends on the excitation intensity. Intertrial time delay should be sufficient to allow for recovery from bleaching.

2. *Spike-triggered averaging of trials with action potentials (refer to Fig. 5).* Normal temporal jitter can be on the order of milliseconds when eliciting action potentials, which can severely affect the averaging of action potential waveforms which have durations on the same order. Time shifts can be easily obtained from the

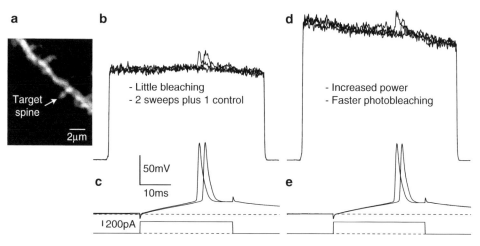

Fig. 4 Single-sweep voltage-sensitive dye (VSD) recordings of backpropagating action potentials in single spines. (**a**) 2-photon fluorescence image of proximal basal dendrite of a cortical pyramidal neuron from a p31 mouse. The spine targeted for all recordings is indicated by the *arrow*. 1,060 nm excitation wavelength with epifluorescence collection only. (**b**) Three sample sweeps from protocol that alternates between action potential (AP) and control trials, with two AP and one control trial shown. (**c**) Somatic, whole-cell membrane potential and current stimuli for AP and control trials. Significant temporal jitter between AP trials can be seen. (**d**) Sample trials with increased excitation power leading to increased bleaching compared to panel (**b**), same scale. (**e**) Somatic electrical recordings

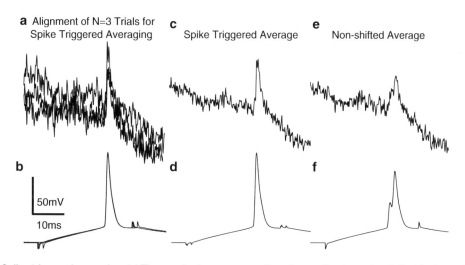

Fig. 5 Spike-triggered averaging. (**a**) Three optical sweeps are aligned according to peaks of electrical waveforms (panel **b**), which are relatively noise-free. The same time shifts necessary for alignment of the electrical waveforms are also applied to the optical recordings. Same data as Fig. 4d. (**c**) Once temporally aligned, trials are averaged for noise reduction, $N=3$ trials. (**d**) Average of aligned electrical waveforms, stimulus onset and offset artifacts appear shifted and attenuated. (**e**) Consequences of jitter-induced misalignment and averaging without time shifting: amplitude is reduced and waveforms are smeared and distorted. (**f**) Simple, unshifted average of electrical waveforms showing the same waveform distortion

peaks of the electrical recordings, which are virtually noise free (Fig. 5b), and can then be used to properly align the fluorescence recordings for averaging and elimination of temporal jitter. With both electrical and optical recordings sampled at the same rate of 0.1 ms, no interpolation or conversion of number of samples to time shift is necessary. Averaging without shifting leads to signal distortion in the fluorescence as well as somatic electrical recordings as demonstrated in Fig. 5e.

3. *Measuring bleaching time constant, sensitivity, and signal-to-noise (refer to Fig. 6).*

(a) Control trials can be averaged to allow for analysis of bleaching and noise. The bleaching time constant is found by fitting a single exponential (Fig. 6a).

(b) Subtracting the exponential fit leaves a waveform consisting of purely noise, which can be quantified in order to measure the signal-to-noise. In theory, the standard deviation can be taken at this point (Fig. 6b, 0.0103, indicated by dashed lines). In practice, low-frequency vibration can contribute unwanted noise. However, this can be estimated with a low-pass filter (Fig. 6b, grey line, 100 Hz cutoff) and subtracted leading to a new estimate of noise (Fig. 6c, 0.093).

(c) Remove bleaching and measure the signal amplitude (Fig. 6d). In theory, the exponential fit from the control trials can be subtracted from the average (spike-triggered) of the AP trials (Fig. 6d, dashed line). However, in practice one must check that the overall level is consistent with the pre-AP data. In the present example it is clearly too high and is adjusted down to fit properly. The estimate of the bleaching time constant is unaffected. After subtraction, the signal amplitude is obtained by measuring the peak fluorescence (Fig. 6e, circle).

(d) Sensitivity is calculated by measuring $\Delta F/F$ in response to a bAP in a spine, where ΔF is the amplitude of the modulation of the fluorescence by the AP and F is the pre-AP baseline fluorescence (with background subtracted, see Subheading 3.2, step 6). We assume that the electrical AP measured in the soma is insignificantly reduced in a proximal dendritic spine. Therefore, dividing the $\Delta F/F$ measured in a proximal dendritic spine by the amplitude of the measured electrical AP provides a calibration of the dye sensitivity in units of $\Delta F/F$mV. We have found this sensitivity to be quite consistent from spine to spine, but not comparable to the voltage sensitivity of the dye in dendritic shafts or somatic membrane regions.

(e) Signal-to-noise is measured by taking ΔF and dividing by the noise found in step (b) above. This is best done on the

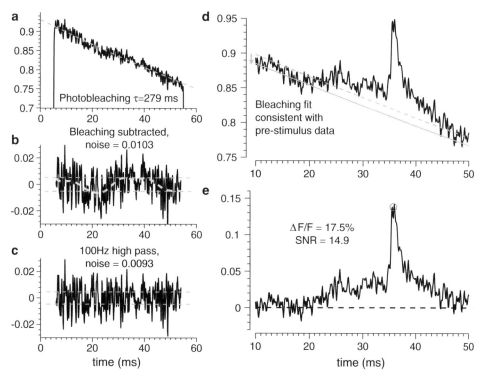

Fig. 6 Analysis of photobleaching, voltage sensitivity, and signal-to-noise. (**a**) Average of three control sweeps with no stimulus with exponential fit, time constant = 279 ms (*dashed line*). (**b**) Noise signal after subtraction of bleaching curve. *Horizontal dashed lines* indicate standard deviation. *Thick grey dashed line* is a low-pass filtered signal used to follow the visible, low-frequency non-stationarity, most likely due to vibration. (**c**) Noise signal after subtraction of the low-frequency modulation with new, slightly reduced standard deviation indicated. (**d**) Spike-triggered average (Fig. 5c) superimposed on bleaching fit from panel (**a**) (*dashed line*). The significant mismatch between the bleaching fit and the pre-stimulus data can be corrected by adjusting the amplitude of the exponential but maintaining the same bleaching time constant (shifted down as indicated by *arrow*). (**e**) Final averaged data with peak intensity indicated along with calculated $\Delta F/F$ and signal-to-noise (for $N = 3$ trials averaged)

averaged data. Estimate the single-sweep signal-to-noise by dividing by $\sqrt{N}$, where N is the number of sweeps or trials.

4. *Examining wavelength-dependence of sensitivity and signal-to-noise.* As noted in the introduction, it is critical to optimize the choice of excitation and emission wavelengths to achieve the best $\Delta F/F$ and S:N. For comparing different wavelengths, laser power is varied in order to match the baseline fluorescence levels; longer wavelengths require greater excitation power. In this way, "noise" is essentially matched between recordings such that signal follows the underlying wavelength-dependent voltage sensitivity (Fig. 7d, e). Despite the need for higher intensities at the edge of the excitation spectrum to produce the same fluorescence intensity, the dye bleaching would not be increased because of the lower absorptivity of the dye at

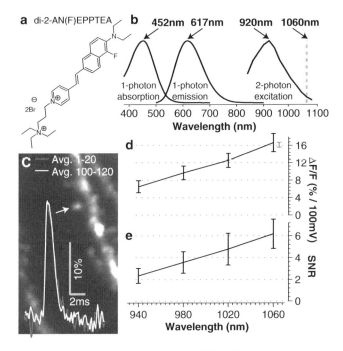

Fig. 7 Stability, voltage sensitivity of di-2-AN(F)EPPTEA. (**a**). Chemical structure of the dye. (**b**) 1-photon absorption and emission spectra in multilamellar lipid vesicles with peak wavelengths shown. 2-photon excitation spectrum measured in a brain slice—excitation power (after objective) was 2.5 mW for all wavelengths. (**c**) Overlapped averages of first and last 20 (of 120) recordings of backpropagating action potentials (bAPs) in a single spine, with no change in signal. (**d**) 2-photon voltage sensitivity as a function of excitation wavelength, measured using bAP amplitudes at proximal spines, normalized by somatic AP amplitude (100 mV somatic). For each of $N = 9$ spines (five cells) all four wavelengths were tested (error bars are SD). Separate point: All proximal spine measurements at 1,060 nm (16.1%, $N = 15$ spines, seven cells, error bar is SEM). (**e**) "Single-sweep" signal-to-noise (SNRs). SNR values are normalized by $\sqrt{N}$ (number of sweeps averaged). Typical laser powers: 1 mW at 940 nm, 4 mW at 1,060 nm. Bleaching is negligible in all instances. Reproduced from (3), with permission from Elsevier

the spectral edge. On the other hand, setting the emission wavelength at the edge of the spectrum wastes many photons reducing S:N; therefore we normally excite at the spectral edge and use emission filters to collect most of the emitted photons. This is a good compromise between maximizing $\Delta F/F$ and maintaining good S:N.

4   Notes

1. A small positive pressure while entering the solution ensures that the tip stays clean and internal solutions are not diluted by the influx of external ACSF solution. The pressure necessary

for patching (~30 mbar, solutions must be filtered) depends on the actual tip size, and must be sufficient to "clear" the tissue as the pipette tip passes through and make a "dimple" on the cell surface, but not too much to damage the tissue and the target cell.

2. The optimal intensity criterion is determined empirically by trial and error based on the ultimate success of the experiment; it should be increased when the resulting signal-to-noise is insufficient, or decreased if signs of toxicity appear.

3. Pipette resistances can be large for the filling electrode, which may result in poor access resistance, but which may improve the cell viability after patch removal. Fill times become longer as a consequence, but this is why highly concentrated dye concentrations are used in the pipette.

4. Changes in appearance such as wrinkling or fading may indicate problems with osmolarity or other issues (pipette vibration is one example).

5. For time efficiency, multiple cells can be filled before going back to repatch the first cell, giving the dye sufficient time to stain processes such as dendrites, which typically requires >30 min. Cell filling can take anywhere from 10 to 40 min.

6. For working at elevated temperatures, it is usually easier to fill multiple cells at room temperature and then increase the temperature before repatching.

Acknowledgements

This work was supported by National Institutes of Health grants No. R01 EB001963 and No. P41 RR013186. We are grateful for the dedication of the people responsible for the synthesis of the voltage-sensitive dyes, Ping Yan and Joseph Wuskell.

References

1. Canepari M, Zecevic D (eds) (2010) Membrane potential imaging in the nervous system: methods and applications. Springer, New York
2. Holthoff K, Zecevic D, Konnerth A (2010) Rapid time course of action potentials in spines and remote dendrites of mouse visual cortex neurons. J Physiol 588(Pt 7):1085–1096
3. Acker C, Yan P, Loew L (2011) Single-voxel recording of voltage transients in dendritic spines. Biophys J 101(2):L11–L13
4. Palmer LM, Stuart GJ (2009) Membrane potential changes in dendritic spines during action potentials and synaptic input. J Neurosci 29(21):6897–6903
5. Peterka DS, Takahashi H, Yuste R (2011) Imaging voltage in neurons. Neuron 69(1):9–21
6. Baker BJ et al (2008) Genetically encoded fluorescent sensors of membrane potential. Brain Cell Biol 36(1–4):53–67
7. Akemann W et al (2010) Imaging brain electric signals with genetically targeted voltage-sensitive fluorescent proteins. Nat Methods 7(8):643–649
8. Tsutsui H et al (2008) Improving membrane voltage measurements using FRET with new fluorescent proteins. Nat Methods 5(8):683–685

9. Kralj JM et al (2011) Electrical spiking in Escherichia coli probed with a fluorescent voltage-indicating protein. Science 333(6040): 345–348

10. Antic SD (2003) Action potentials in basal and oblique dendrites of rat neocortical pyramidal neurons. J Physiol 550(1):35–50

11. Judkewitz B et al (2009) Targeted single-cell electroporation of mammalian neurons in vivo. Nat Protoc 4(6):862–869

12. Pologruto TA, Sabatini BL, Svoboda K (2003) ScanImage: flexible software for operating laser scanning microscopes. Biomed Eng Online 2:13

13. Acker CD, Antic SD (2009) Quantitative assessment of the distributions of membrane conductances involved in action potential backpropagation along basal dendrites. J Neurophysiol 101(3):1524–1541

14. Hamill OP et al (1981) Improved patch-clamp techniques for high-resolution current recording from cells and cell-free membrane patches. Pflugers Arch 391(2): 85–100

Chapter 12

Characterization and Validation of Fluorescent Receptor Ligands: A Case Study of the Ionotropic Serotonin Receptor

Ruud Hovius

Abstract

The application of fluorescent receptor ligands has become widespread, incited by two important reasons. "Seeing is believing"—it is possible to visualize in real time in live cells ligand–receptor interactions, and to locate the receptors with subcellular precision allowing one to follow, e.g., internalization of the ligand–receptor complex. The high sensitivity of photon detection permits observation of on the one hand receptor–ligand interactions on cells with low, native receptor abundance, and on the other of individual fluorophores unveiling the stochastic properties of single ligand–receptor complexes.

The major bottlenecks that impede extensive use of fluorescent ligands are due to possible dramatic changes of the pharmacological properties of a ligand upon chemical modification and fluorophore conjugation, aggravated by the observation that different fluorophores can provoke very dissimilar effects. This makes it virtually impossible to predict beforehand which labelling strategy to use to produce a fluorescent ligand with the desired qualities.

Here, we focus on the design, synthesis, and evaluation of a high-affinity fluorescent antagonist for the ionotropic serotonin type-3 receptor.

Key words 5-HT3 receptor, Antagonist, Fluorophore conjugation, Ligand–receptor interaction, Affinity, Efficacy, Fluorescence anisotropy, Fluorescence intensity

1 Introduction

Receptor–ligand interactions have been, and still are being, investigated using radioligands, whose major advantage is that the label can be introduced by isotope replacement, or by the introduction of a small group carrying a radioisotope, thus with minimal perturbation of the molecular properties. However, the use of radioisotope-labelled ligands presents also certain drawbacks, related on the one hand to safety and waste disposal issues, and on the other to the low sensitivity prohibiting adequate temporal resolution and real-time in vivo applications.

Fluorescent ligands circumvent these problems, as fluorescence detection is inherently far more sensitive than isotope decay

Matthew R. Banghart (ed.), *Chemical Neurobiology: Methods and Protocols*, Methods in Molecular Biology, vol. 995,
DOI 10.1007/978-1-62703-345-9_12, © Springer Science+Business Media New York 2013

detection (~10^6 times), and due to its noninvasive character permits high-resolution real-time live cell imaging. As intrinsically fluorescent ligands are rare, and in general their fluorescence properties are weak, one prefers to conjugate bright organic fluorophores to receptor–ligands. Literature presents excellent reviews (1–5).

Before setting out to synthesize a new fluorescent receptor–ligand, one should evaluate which information is desire and consider whether this would be possible to achieve with the available instrument(s). Also one should consider whether the fluorescent ligand should be an agonist or an antagonist. Importantly, agonists activate receptors to induce cellular responses, which has several consequences. (1) The *affinities* of agonists for their cognate receptors cannot be determined as the reactions of ligand-binding and cellular response are coupled; thus the measured *apparent affinity* does not equal the actual *affinity* (6). However, one can determine the *efficacy* of a fluorescent agonist to induce a cellular response, e.g., changes in intracellular calcium concentrations. (2) The location of receptors is prone to change due to, e.g., feedback mechanisms leading to receptor internalization. This, however, could be the aim of an investigation. The use of fluorescent antagonists, i.e., molecules that do not change the activation state of a receptor, might be preferred to circumvent the problems mentioned above.

Discrimination of receptor-bound from free ligands is important, as both are fluorescent. The (apparent) affinity of a fluorescent ligand and the amount of receptor protein available determine whether one can perform homogeneous (mix and measure) assays or has to resort to heterogeneous (mix, separate, then measure) assays. For the latter assays, one has to take into account the dissociation of the receptor–ligand complex and the time needed for separation and measurement. In general, the limiting amount of receptor available imposes that only low concentrations of this protein can be achieved. The higher the affinity of a fluorescent ligand for its receptor, the lower the ligand concentration needed to achieve a given receptor occupancy, resulting in larger fraction of bound ligand, increasing the experiment's sensitivity that depends on the ratio bound/free.

The design of a new fluorescent ligand is not a *sinecure*; it has to result in a combination of high affinity and high selectivity. It is very helpful to explore the literature for compounds that have been used for ligand-affinity chromatography purification of receptor proteins, or for structural information on the pharmacophore and/or the receptor of interest. If no such information is available, one has to resort to trial-and-error modification of available ligands to introduce a fluorophore either directly or via a linker. Of relevance here is a recent publication on the development of a fluorescent serotonin type-3 receptor (5-HT3R) antagonist based on granisetron (7).

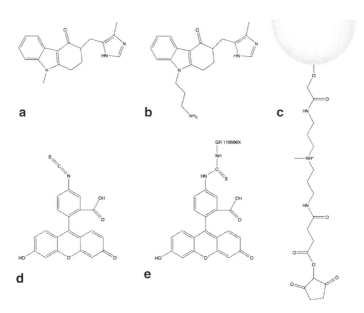

Fig. 1 Structure of (**a**) GR-67330, (**b**) GR-119566X, (**c**) the spacer of Affigel 15 used to link GR-119566X to the chromatography resin illustrated by the *half sphere*, (**d**) fluorescein isothiocyanate (FITC), and (**e**) GR-Flu (=GR-186741X)

Here, we prepare a fluorescent antagonist for the 5-HT3R. The high-affinity antagonist GR-67330 ($K_d = 110$ pM (8)) was modified by replacing the 9-methyl by a 9-propylamine yielding GR-119566X (GR-H) (Fig. 1). This compound preserved the high affinity ($K_d = 65$ pM (9)) and has been coupled via the introduced primary amino group to a chromatography resin enabling the affinity purification of the 5-HT3R (8, 9). This propylamine spacer is thus an obvious choice for the attachment of fluorophores, in this chapter fluorescein yielding GR-186741X, hereafter called GR-Flu (10).

2 Materials

2.1 Preparation of GR-Flu

1. GR-119566X (1,2,3,9-tetrahydro-3-((5-methyl-1H-imidazol-4-yl)methyl)-9-(3-aminopropyl)-4H-carbazol-4-one, Glaxo Institute of Molecular Biology, see Note 1).

2. Fluorescein isothiocyanate (FITC, Sigma, see Note 2).

3. Anhydrous dimethylformamide (DMF) and dimethylsulfoxide (DMSO) on molecular sieve (Sigma, see Note 3).

4. Di-isopropylethylamine (DIPEA, Sigma).

5. Screw-cap polypropylene (Sarstedt, 1.5 ml).

6. Silica G60 F_{254} on 20×20 cm aluminium foil (Merck), cut to approx 5×7 cm.

7. 0.25 or 0.5 mm thick silica G60 HPTLC on glass thin layerchromatography (TLC) plates of 10×20 cm (Merck, see Note 4).

8. Chromatography tank from glass.

9. TLC eluent composed of methanol:ammonia (95:5 v/v).

10. UV lamp, with UV mask for eye protection.

11. Plastic 10 µl pipette tip and a 10 µl pipette.

12. Oil pump and vacuum desiccator.

13. Razor blade and dust mask.

14. Weighting paper.

15. Round-bottom flask and a G4 glass filter.

16. Methanol.

17. Water pump.

18. Speed vac linked to an oil pump equipped with a trap for organic solvents.

2.2 Mass Spectrometry (MS) Analysis

1. TSQ 7000 mass spectrometer using electron spray ionisation (Thermo Fisher).

2. Acetonitrile and acetic acid.

3. Tabletop centrifuge.

4. Injection syringe.

2.3 Receptor Expression and Receptor Preparations

1. HEK 293T cells (ATCC).

2. DMEM/F-12 + GlutaMax (Gibco) supplemented with 10% newborn calf serum (Sigma), 100 units/ml penicillin, and 100 µg/ml streptomycin from 100-fold dilution of PenStrep solution (Gibco, see Note 5).

3. Trypsin (0.05%)/EDTA (0.53 mM) solution (Gibco).

4. Incubator for a 37°C humidified 5% CO_2 atmosphere.

5. Calcium phosphate transfection solutions A (250 mM $CaCl_2$) and B (1.4 mM NaH_2PO_4 or Na_2HPO_4, 140 mM NaCl, 50 mM Hepes), adjusted to pH 7.05 with NaOH or HCl (see Note 6).

6. Plasmid for mammalian expression of the 5-HT3R, e.g., pEAK8 (EdgeBio).

7. T25 culture flasks and 6-well plates (TPP).

8. Phosphate-buffered saline (PBS, Sigma) with 5 mM EDTA pH 7.4.

9. 10 mM Hepes, adjusted to pH 7.4 with HCl.

10. Ultra-turrax T25 cell homogenizer (IKA).

11. Centrifuge tubes.

12. 10% dodecylnonaethyleneglycol ($C_{12}E_9$) in vials under argon (Anatrace).

13. Complete mini protease inhibitor cocktail tablets (Roche).

2.4 Radiologand Binding (See Note 7)

1. (^{3}H)-GR65630 (PerkinElmer), with a specific radioactivity of about 1.5×10^{17} decays per minute (dpm)/mole, store at –20°C.

2. 10 mM quipazine dimaleate (Tocris), store aliquots at –20°C.

3. GF-B glass filters, 25 mm diameter (Whatman).

4. 0.5% polyethyleneimine (PEI), 100-fold dilution of 50% PEI (Aldrich).

5. 10 mM Hepes (adjusted to pH 7.4 with HCl), store at 4°C.

6. Filtration manifold (Millipore) for 12 filters, connected to a vacuum pump.

7. Ultima-Gold scintillation liquid (Perkin Elmer).

8. Scintillation vials holding 6 ml (Packard).

9. Liquid scintillation counter Tri-Carb 3100-TR (Packard).

2.5 Fluorescence Intensity and Polarization

1. Spex Fluorolog II fluorimeter (Instrument SA) in L-format, equipped with a magnetic stirrer and a cuvette holder, thermostated by a circulating water bath.

2. Home-made polarizer set-up with large 15×15 mm Glan Taylor polarizers (Halbo Optics) apposed to the cuvette holder to maximize light throughput.

3. 10×4 mm quartz cuvette with compartment for stirring bar (Hellma).

4. 10 mM Hepes, 0.4 mM $C_{12}E_9$, pH 7.4.

3 Methods

The creation of fluorescent receptor–ligands with favorable properties is a process of educated trial and error. Major decision points are (1) the chemistry to attach a handle to the pharmacophore to allow probe conjugation, and (2) the choice of the fluorophore. The importance of the latter is often not appreciated; however, the following examples are instructive:

– GR-H conjugated with either fluorescein or Cy5 yields ligands with a 10- or 1,000-fold decreased affinity with respect to the parent component, respectively (11).

– The related peptide-toxins α-conotoxin GI and α-conotoxin IMI target selectively the muscle-type and neuronal α7 acetylcholine receptors, respectively. Upon conjugation with Atto-647N, the former remains a specific probe of its receptor, whereas the latter has become a mitochondrial marker (Hovius, unpublished).

Careful functional analysis of affinity, efficacy, and specificity of the produced labelled ligand is important, before it can be generally applied to investigate its cognate receptor.

3.1 Preparation of GR-Flu (See Note 8)

3.1.1 Conjugation of Fluorescein to GR-H

Polypropylene screw-cap reaction tubes are used in steps 1–3.

1. Dissolve 0.9 mg GR-H in 90 μl DMF, 30 μl DMSO, and 5 μl DIPEA.

2. Dissolve 1.2 mg FITC in 100 μl DMF.

3. Combine 115 μl GR-H (2.5 μmol) and 90 μl FITC (2.75 μmol) solutions, mix, and incubate in the dark at room temperature. The remainder of the GR-H and FITC solutions are used as standards for analysis.

4. Analyze the reaction progress by TLC. Draw, with a soft pencil caring not to damage the silica layer, a horizontal line at 1 cm above the bottom of a TLC plate. Apply next to each other as a 3–5 mm stripe ~3 μl of diluted aliquots from, respectively, GR-H, reaction mixture, and FITC. Let dry for 15 min in the dark. Meanwhile, fill a chromatography tank, or a jam jar, with about 5 mm with eluent; close the tank and shake to achieve vapor equilibration (see Note 9). Immerse the bottom of the TLC plate in the eluent and close the tank. Develop the plate until the eluent has almost reached the top of the plate. Remove the plate from the tank, and mark the eluent front with a pencil.

Fluorescent components are directly visible. Nonfluorescent ones are revealed as dark spots upon illumination with UV-light, as they block fluorescence of the F_{254} fluorescence indicator present in the TLC plate.

An example is shown in Fig. 2a, demonstrating that the reaction has progressed to completion, as there is no GR-H left.

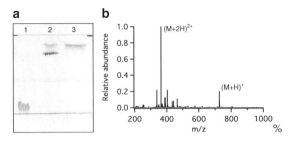

Fig. 2 Analysis of GR-Flu. (**a**) TLC separation of (*1*) GR-H, (*2*) GR-Flu synthesis reaction mixture, and (*3*) FITC, developed with methanol:ammonia (95:5 v/v). The *horizontal lines* at the *bottom* and *top* indicate the positions of application of the samples and final eluent front, respectively. (**b**) Positive-ion mass spectrum of GR-Flu after purification. Indicated are the $(M+H)^+$ and $(M+2H)^{2+}$ peaks at *m/z* = 726.4 and *m/z* = 364.3, respectively

The product GR-Flu has now to be separated from remaining FITC and possibly remaining traces of GR-H.

1. Draw with a soft pencil an 18 cm long horizontal line at 1 cm from the bottom of a preparative HPTLC plate, starting and stopping at 1 cm from the side of the plate. Often separation of components at the ends is less good.

2. Carefully and slowly apply the reaction mix as a fine stripe along the pencil line using a 10 μl pipette tip by gently touching the silica layer. Do not use 1 cm at either extreme of the line. Let dry the plate in the dark for 30 min, e.g., in a drawer, followed by 1 h in a desiccator linked to an oil pump, to remove solvent and base (see Note 10).

3. Prepare a large chromatography tank with 5 mm TLC eluent, and placing a piece of filter paper of about the size of the TLC plate against one of the walls; close the tank and shake to wet the filter paper to efficiently saturate the air with eluent ensuring optimal separation. Insert the dried TLC plate, posing it against the free wall. Be sure that the sample is not emersed in the eluent. Envelop the tank with a cardboard box or alu-foil to shield the sample from light. Develop the plate until the eluent has almost reached the top of the plate. Remove the plate from the tank, and mark with the pencil the eluent front. The chromatogram should resemble the TLC made in Subheading 3.1.1 (Fig. 2a).

4. Put on a dust mask. Before the solvent has completely dried (see Note 11), remove the silica around the product, the lowest fluorescent band, using a razor blade (Fig. 2a). Then scratch off the silica band containing the product onto a weighing paper facilitating its transfer onto a glass filter, and let it dry for 15 min in the dark.

5. Extract the product from the silica by sequentially applying 0.5 ml methanol and gently mixing the slurry with a small glass rod. Collect the filtrate in a flask; when needed apply pressure to accelerate filtration. Repeat until no more product is extracted from the silica.

6. Determine the concentration of GR-Flu in the final combined filtrate and the yield of the reaction by measuring the absorbance of a diluted GR-Flu solution in 0.1 M NaOH, using an extinction coefficient of 76,000 M^{-1} cm^{-1} for the absorbance band of fluorescein at 488 nm. The yield of the synthesis after purification is in general 40–60%, thus about 1.25 mmol (~1 mg) of GR-Flu. This does not seem a lot; however, one can perform about 10^9 experiments when using 200 μl of a 15 nM solution for a sample in a standard 96-well plate or an 8-well used for microscopy!

7. Aliquot the solution in screw-cap vials, and remove the solvent using a speed-vac, i.e., centrifuge connected to a lyophilizer. Remove the final traces of solvent by placing the vials in a vacuum desiccator connected to an oil pump. During this procedure, do shield the product from light. Tightly close the vials and store at –80°C. Thus stored GR-Flu is stable for years.

3.2 Chemical Characterization

The identity of a produced compound is most commonly demonstrated by combined analysis with nuclear magnetic resonance (NMR) or mass spectrometry (MS). However, the conjugation reaction performed here is a simple one with hardly any possible side reactions, and by chromatography it has been shown that the product is different from the adducts. Therefore, analysis by MS alone will be sufficient. Here, we use a TSQ7000 (Thermo Fisher) spectrometer using electron spray ionization, employing as carrier phase acetonitrile/H_2O/acetic acid (50/50/1).

1. Prepare a solution containing about 5 µg GR-Flu in 50 µl MeOH.

2. Rinse seven times with methanol the syringes for sample injection and injection port washing, touching the tip of the needle to a tissue while ejecting.

3. Centrifuge the sample before injection to remove possible particles, and wash the injection port thoroughly.

4. Inject 10 µl of sample and start acquisition of a positive ion mass spectrum in the 200–1,000 m/z range.

5. Analyze the mass spectrum (Fig. 2b): Observed are the singly $(M+H)^+$ and doubly $(M+2H)^{2+}$ charged molecular ions at $m/z=726.4$ and 364.3, respectively. This confirms that the product is indeed GR-Flu.

3.3 Receptor Expression and Preparations

The amounts indicated are for a single well of a 6-well plate, which has a surface of ~8 cm^2.

1. HEK293 cells are cultured in DMEM-medium at 37°C in a humidified 5% CO_2 atmosphere. To maintain the culture, the cells are subcultured 1–10 when reaching 80–90% confluency using trypsin/EDTA solution.

2. Transfer 150,000 cells in 2 ml medium for each well in a 6-well plate.

3. Transfect cells 16 h later. Mix 2.5 µg DNA with 200 µl transfection solution A. Add 200 µl transfection solution B and mix for 15 s by gently pipetting up and down. At 60 s after addition of solution B, add the DNA drop-wise to the cells. Exchange the medium after 4 h (see Note 12).

4. Harvest the cells 36–48 h after transfection using 1 ml PBS/EDTA, and transfer to a centrifuge tube. Isolate cells by centrifuging for 5 min at $500 \times g$.

5. Suspend the cell pellet in 2 ml Hepes buffer, and homogenize for 10 s with an ultra-turrax. Isolate membranes by centrifugation for 10 min at $27,000 \times g$, and resuspend in 2 ml Hepes buffer to yield the membrane preparation, which can be stored at $-80°C$ until use (see Note 13).

6. Solubilize the 5-HT3R from the membrane preparation by adding $C_{12}E_9$ to 0.4 mM, incubating for 60 min on ice, and recuperating the supernatant after centrifugation for 1 h at $100,000 \times g$. In general the efficiency of solubilization is 50–80%, as evaluated by radioligand binding. Store aliquots at $-80°C$ until further use.

3.4 Determination of the Affinity of GR-Flu by Radioligand Competition Assay

The functional characterization of a compound assumed to interact with the 5-HT3R, a ligand-gated ion channel, should evaluate both its affinity and efficacy by radioligand competition assays and electrophysiology, respectively.

The amount of ligand–receptor complex $RL*$ formed upon binding of a radioligand $L*$ to a receptor R (Eq. 1) depends on the concentration of total receptor R_T and free radioligand $L*$, as well as the dissociation constant K_D^* describing this equilibrium (Eqs. 1 and 2), according to the Langmuir equation (Eq. 3) (12), and assuming the absence of ligand depletion $L*$ equals the total ligand concentration $[L_T^*]$ (see Note 14):

$$R + L^* \rightleftarrows RL^* \qquad (1)$$

$$K_D^* = \frac{[L^*] \cdot [R]}{[RL^*]} \qquad (2)$$

$$[RL^*] = \frac{[R_T]}{1 + \dfrac{K_D^*}{[L^*]}}. \qquad (3)$$

In the presence of a competitive inhibitor I that binds to the same site as $L*$, the formation of $RL*$ depends also on $[I]$ and the dissociation constant K_i of the binding of I to R (Eqs. 4–6). Knowing K_D^*, one can experimentally determine K_i by performing a radioligand competition assay, quantifying $RL*$ with constant $[L*]$ and $[R_T]$, and varying $[I]$. K_i can be evaluated by fitting the data to Eq. 6 that yields the IC_{50}, the $[I]$ where $[RL*]$ equals 50% of $[RL*]_{[I]=0}$ in the absence of I, and applying the Cheng–Prusoff equation (13) (Eq. 7):

$$R + I \rightleftarrows RI \qquad (4)$$

$$K_i = \frac{[I] \cdot [R]}{[RI]} \qquad (5)$$

$$[RL^*]_{[I]} = \frac{[RL^*]_{[I]=0}}{1 + \dfrac{\mathrm{IC}_{50}}{[I]}} \qquad (6)$$

$$K_i = \frac{\mathrm{IC}_{50}}{1 + \dfrac{[L^*]}{K_D^*}}. \qquad (7)$$

A single 6-well holds about 500,000 cells, each expressing ~500,000 5-HT3R, in total 0.4 pmol of receptor. Considering the specific radioactivity of the radioligand, this corresponds to 60,000 dpm of ligand-binding sites. Determination of K_i of GR-Flu through a competition binding experiment needs ten concentrations of GR-Flu to be evaluated in triplicate with about 1,500 dpm binding sites per sample.

This protocol for GR-Flu is valid for other ligands upon adaption of the concentrations of competing ligand to the range from $0.01 \times K_i$ to $100 \times K_i$.

1. Prepare ten 500 µl solutions of GR-Flu in Hepes buffer ranging from 0.05 to 300 nM, e.g., 300, 100, 30, 10, 5, 2, 1, 0.5, 0.15, and 0.05 nM.

2. Prepare a (^{3}H)-GR-65630 stock solution by mixing 1 µl (^{3}H)-GR-65630 with 1,200 µl Hepes buffer; this will yield a concentration of approx 20 nM.

3. Make 1 ml of 10 µM quipazine solution in Hepes buffer.

4. Prepare 12 times three 1.5-ml reaction tubes, each containing:

	Tubes	Volume	Remarks
Hepes buffer	All	820 µl	
Radioligand	All	30 µl	
Competing ligand	1–3	100 µl Hepes buffer	100% binding
	4–33	100 µl GR-Flu dilutions	Competition
	34–36	100 µl quipazine	0% binding
5-HT3R membranes	All	50 µl	Pipette in tube lid

5. Start the experiment by closing the lid and thoroughly mixing the entire contents; incubate for 1 h at RT.

6. Meanwhile, submerge glass filters in PEI solution. Activate the vacuum pump, and place 12 filters on the filtration manifold.

7. Recuperate receptor-bound ligand by applying 800 µl per sample to a filter, and quickly washing twice with 3 ml of ice-cold Hepes buffer. Do this for 12 samples one after the other. Transfer the filters to scintillation vials, add 4 ml liquid

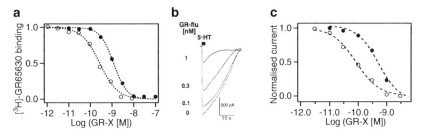

Fig. 3 Analysis of the functional properties of GR-Flu. (**a**) Radioligand binding competition assay to evaluate the affinity of GR-H (*open circle*) and GR-Flu (*filled circle*). (**b**) Whole-cell current evoked by application of 30 μM serotonin (5-HT) in the absence and presence of increasing concentrations of GR-Flu. (**c**) Inhibition of serotonin evoked 5-HT3R currents by increasing concentrations of GR-H (*open circle*) and GR-Flu (*filled circle*). Adapted from (10)

scintillation liquid, close the lids, and vortex. Repeat until all 36 samples have been treated.

8. Prepare three scintillation vials with each 30 μl of radioligand and 4 ml scintillation liquid to determine the radioligand concentration [L_T^*].

9. Load the samples in the racks of the counter and start an appropriate program from tritium yielding dpm values. It takes time for the radioligand to mix with the scintillation liquid; thus repeat the whole counting sequence until the counts have become stable.

10. Plot the average ± standard deviation of bound ligand vs. GR-Flu concentration (samples 4–33), including the value $[RL^*]_{[I]=0}$ of the samples without GR-Flu (samples 0–3) and with quipazine (samples 34–36) that correspond to complete inhibition of specific radioligand binding $[RL^*]_{[I]=\infty}$. Fit the data to Eq. 12.6 to evaluate IC$_{50}$ and use Eq. 7 to obtain the dissociation constant K_i of GR-Flu. Results of an experiment comparing GR-H with GR-Flu gave IC$_{50}$-values of binding of 0.27 ± 0.03 nM and 1.19 ± 0.05 nM (Fig. 3a), respectively, corresponding to K_i of 60 ± 10 pM and 270 ± 20 pM.

Electrophysiological experiments have been performed to evaluate the effect of GR-H-based compounds on the activation of 5-HT3R. A description of this method is not in the scope of the present chapter; however, the results are briefly presented (Fig. 3b, c). The application of serotonin (5-HT) to 5-HT3R-expressing cells induced a robust transmembrane current, which is progressively diminished upon the inclusion of increasing concentrations of GR-H or GR-Flu (Fig. 3c), demonstrating that these compounds are 5-HT3R-antagonists. Evaluation of the peak currents as function of concentration yields the IC$_{50}$-values of channel activation for GR-H or GR-Flu of 85 ± 6 pM and 460 ± 40 pM, respectively (Fig. 3c).

3.5 Determination of the Affinity of GR-Flu by Fluorescence Measurements

The binding of a small ligand to a large receptor protein can be evaluated from the decrease in mobility of the ligand upon receptor binding. The rotational mobility is quantified from the anisotropy of the emitted fluorescence, whereas the translational mobility can be determined by, e.g., fluorescence correlation spectroscopy. The latter method is not discussed here; the reader is referred to literature (e.g., (14)).

The microenvironment of a chromophore affects its fluorescence properties. For instance, fluorescein is highly fluorescent at basic pH, whereas under acidic conditions its fluorescence intensity is strongly reduced. Moreover, a decrease of the polarity of the medium diminishes the fluorescence intensity considerably with a concomitant bathochrome shift of the emission. In the case of GR-Flu, the fluorescence intensity decreases strongly upon receptor binding (10) (Fig. 4b).

Detergent-solubilized receptors are used for these experiments, as receptor-containing membrane fragments scatter light, inducing artifacts in the determination of the anisotropy of the emitted fluorescence.

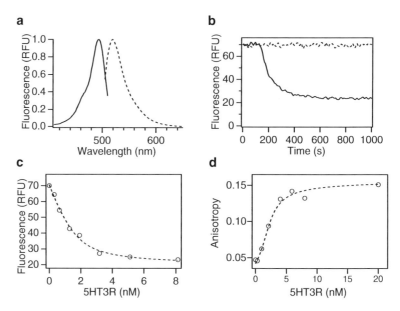

Fig. 4 Fluorescence quantification of the binding of GR-Flu to 5-HT3R. (**a**) Fluorescence excitation (*solid line*) and emission (*dashed line*) spectra of GR-Flu. (**b**) Fluorescence intensity time trace of a 2 nM GR-Flu solution without (*solid line*) or with 1 μM quipazine (*dashed line*) to which at 125 s 5-HT3R is added to 8 nM. The difference in fluorescence intensity at equilibrium corresponds to ΔF (Eq. 8). (**c**) The change in fluorescence intensity F_i with 5-HT3R concentration. Fitting (*dashed line*) with Eq. 8 modified to account for ligand depletion, yielding $K_D = 0.24 \pm 0.08$ nM. (**d**) The anisotropy r of 2 nM GR-Flu upon the addition of 5-HT3R to the indicated concentrations. A $K_D = 0.22 \pm 0.04$ nM was obtained upon application of Eqs. 12–14. Adapted from (10)

1. The fluorescence properties of GR-Flu are investigated on a Spex Fluorolog II using 1.0 and 1.35 nm band-passes for excitation and emission, respectively. The temperature was maintained at 20°C using the circulating water bath. 10×4 mm cuvettes filled with 1 ml solution and constantly stirred with a magnetic stirrer bar are used (see Note 15).

2. Record excitation (excitation = 410–510 nm and emission = 520 nm) and emission (excitation = 488 nm and emission = 505–650 nm) spectra (1 nm step and 0.5-s integration time) of 10 mM Hepes, 0.4 mM $C_{12}E_9$ (pH 7.4); these are blank spectra that have to be subtracted from sample spectra. Note the small peaks at approx 440 nm and 592 nm in the excitation and emission spectra, respectively; these are due to the Raman vibration of water. Perform the same experiment with a solution of 2 nM GR-Flu, and subtract the blank spectra (Fig. 4a) (see Note 16).

3. Prepare a cuvette with 1 ml of 2 nM GR-Flu in 10 mM Hepes, 0.4 mM $C_{12}E_9$ (pH 7.4). Start a time scan exciting at 495 nm and acquiring the fluorescence at 520 nm with 6-s intervals and 1-s integration time. A stable fluorescence signal should be observed. At about 125 s, add solubilized 5-HT3R to approx 8 nM, mix rapidly without producing air bubbles, and continue acquisition to 1,000 s (Fig. 4b). Repeat the same experiment including 1 μM quipazine in the buffer; the excess of quipazine prevents GR-Flu binding to the 5-HT3R (Fig. 4b).

4. Repeat step 3 with receptor concentrations between 0 and 8 nM.

5. Report in a graph for each receptor concentration $[R]_i$ the final equilibrium fluorescence intensity F_i (Fig. 4c), and evaluate K_D using

$$F_i = F_0 + \frac{\Delta F_{max}}{1 + \dfrac{K_D}{[R]_i}}, \tag{8}$$

where F_0 is the fluorescence intensity in the absence of receptor, and ΔF_{max} the maximal fluorescence change attainable.

The affinity ($= 1/K_i$) of unlabeled compounds can be determined by competition binding experiments, in analogy to the radioligand competition assay, by including different concentrations of competitors in the GR-Flu solution and quantifying fluorescence intensity change upon the addition of a constant receptor concentration. K_i can be evaluated with Eqs. 6 and 7 adapted in a similar manner as Eq. 8 from Eq. 3.

The fluorescence emitted by chromophores in solution upon excitation with linearly polarized light is not uniformly distributed in all directions, which can be described in terms of the anisotropy r, or polarization p, of this fluorescence (see Note 17). This anisotropy is quantified by excitation of a sample with vertically polarized light, and detecting through vertically or horizontally oriented polarizers the intensities of the emitted fluorescence I_{VV} and I_{VH}, respectively. The anisotropy r is calculated using Eq. 9, where G is an instrumental factor correcting for the different detection efficiencies of horizontal I_{HH} and vertical I_{HV} polarized fluorescence upon excitation with horizontally polarized light Eq. 10. The anisotropy r can be converted in polarization p using Eq. 11. Anisotropy is preferably used, especially in cases where several fluorescence emitting species are present, as for receptor–ligand interactions where the anisotropy of free r_{Free} and bound r_{Bound} ligands are different. In this case, the measured anisotropy r equals the sum of the anisotropy of the free and bound species weighted by their mole fractions f_{Free} and f_{Bound} and by the ratio of the molecular fluorescence intensity $Q = FI_{Bound}/FI_{Free}$ (Eq. 12), which equals ~0.3 for GR-Flu (see also Fig. 4c):

$$r = \frac{I_{VV} - G \cdot I_{VH}}{I_{VV} + 2 \cdot G \cdot I_{VH}} \tag{9}$$

$$G = \frac{I_{HV}}{I_{HH}} \tag{10}$$

$$p = \frac{3r}{2 + r} \tag{11}$$

$$r = \frac{f_{Free} \cdot r_{Free} + f_{Bound} \cdot r_{Bound} \cdot Q}{f_{Free} + f_{Bound} \cdot Q}. \tag{12}$$

1. The instrument is as described in Subheading 3.5.1, except that the emission bandwidth was increased to 1.5–2 nm, and that polarizers were positioned in the excitation and emission light paths. Excitation is at 495 nm, and the fluorescence emission is acquired at 520 nm.

2. Acquire for 10 mM Hepes, 0.4 mM $C_{12}E_9$ (pH 7.4) the background fluorescence intensities I_{HH}, I_{HV}, I_{VV}, and I_{VH}. Perform the same measurements upon the addition of GR-Flu to 2 nM, and correct the obtained values for buffer background. Calculate the G-factor (Eq. 10), and using Eq. 9 the anisotropy r_{Free} of GR-Flu free in solution.

3. Determine the anisotropy r of the emitted fluorescence of solutions of 2 nM GR-Flu in 10 mM Hepes, 0.4 mM $C_{12}E_9$ (pH 7.4) in the presence of 5-HT3R at concentrations ranging from 1 to 20 nM. At the highest 5-HT3R concentration virtually all GR-Flu is receptor bound, and the anisotropy measured corresponds to r_{Bound} of the fully bound ligand.

4. Perform control experiments in the presence of 1 μM quipazine.

5. Report the measured anisotropy r vs. 5-HT3R concentration (Fig. 4d), and evaluate the fraction of receptor-bound f_{Bound} ligand (Eq. 12.13) and K_D (Eq. 14):

$$f_{Bound} = \frac{r - r_{Free}}{Q(r_{Bound} - r) + (r - r_{Free})} = \frac{[RL^*]}{[L_T]} \qquad (13)$$

$$K_D = \frac{[L^*][R]}{[RL^*]} = \frac{[L^*_T - RL^*][R_T - RL^*]}{[RL^*]}. \qquad (14)$$

4 Notes

1. Finding molecules is aided significantly by search engines like SciFinder, with which one can search by, e.g., name, CAS code, or structure. The program gives scientific information, known reactions, and commercial sources. Companies that offer many receptor–ligands are, e.g., Tocris, RBI, Maybridge, Toronto Research Chemicals, and Bachem. Noncommercial sources of interest are NIMH Chemical Synthesis and Drug Supply Program (http://nimh-repository.rti.org/), ChemBioNet (http://www.chembionet.info/), and pharmaceutical companies, especially for publish components.

2. Fluorophores with different reactive groups, absorbance and emission wavelengths, chemical structures, physicochemical properties, and prizes are available from, e.g., Atto-tec, Clontech/Molecular Probes, Dyomics, GE Healthcare, TRC, and Biotium.

3. Inform yourself of potential dangers of chemicals and how to handle them properly before starting. Check the R & S numbers.

4. TLC is a cheap and rapid method to rapidly evaluate progress of chemical reactions and to test different eluents in parallel. Preparative separation of 1–2 mg amounts is readily done with 20 cm large plates covered with 0.25–0.5 mm silica. Possible drawbacks of preparative TLC are limitations in separating power and in capacity. Also, it is possible that some silica from the stationary phase dissolves in the solvent used for extraction. This can be prevented by using water-free solvents. Alternatively, one could use a high-performance liquid chromatography (HPLC) instrument equipped with a C18-reversed-phase column using a gradient of 5–100% acetonitrile/0.1% (v/v) TFA in water/0.1% (v/v) TFA. However, this requires a rather consequent financial investment.

5. Antibiotics are not necessary, but prevent to some extent bacterial infection.

6. Transfection of mammalian cells can be done by many different means. The calcium phosphate method is very economical and works well. Alternative methods like lipofection work as well and might be more convenient, but are more costly (see Note 12).

7. Please do consult national and local authorities for regulations concerning working permits and conditions, and waste disposal issues.

8. The method described to produce GR-Flu works to synthesize other GR-H conjugates. Moreover, the method applies to fluorophore labelling of other receptor ligands that present a thiol- (choose, e.g., maleimide-activated fluorophores) or primary amino-group, modifying solvents to account for differences in solubility. See also (1–5).

9. Do have a look during the chromatography to see if there is still enough eluent; especially preparative plates might need a refill of eluent.

10. Removal of solvent and bases or acids of applied samples is essential for a good separation. Especially in the case of solvents with high boiling points like DMF and DMSO, or bases like DIPEA, it is important to make sure that it is completely removed—the nose is a sensitive detector (be careful).

11. Work in a chemical hood. Dry silica is crisp, and scratching will provoke quite some silica dust and large chunks to jump away. Scratching from a slightly "wet" plate prevents these problems.

12. Timing is rather critical. Upon mixing the phosphate and calcium solutions crystals will start to form, and their morphology changes with time. The 1-min interval has been shown to yield the best transfection efficiencies. In case that many samples have to be transfected in parallel, lipofection with, e.g., PEI might be more convenient, as time is less critical.

13. Frozen membrane preparations are stable for months. Depending on the receptor investigated one might need to include protease inhibitor mixtures in all buffers for membrane preparation and receptor solubilization. Commercial cocktails are convenient. Absence of protease inhibitors does not affect the ligand-binding activity of the 5-HT3R.

14. Depletion of reactants can be regarded negligible if less the bound species represents less than 10%. If this is not the case, one should take into consideration that $[L*] = [L*_T] - [RL*]$ and $[R] = [R_T] - [RL*]$.

15. Cleanliness of optical cuvettes is important. Rinsing with incubation buffer alone will not clean them properly, but might be sufficient. Thorough cleaning can be achieved using more aggressive (be careful) solutions like Hellmanex (contains detergents and NTA; to be avoided when working with His-tagged

proteins, Hellma) or nitric acid. Also to dry cuvettes do not use standard paper as this might scratch the cuvette's windows. Rather rinse with ethanol and apply a stream of nitrogen.

16. If the fluorescence signal is not large enough and rather noisy, increase the width of the emission slits, if the needed wavelength resolution permits. If this is not sufficient, increase the integration time or the width of the excitation slits, if the time resolution or the stability of the fluorophore permits. As a last resort, one has to increase the concentration of the fluorescent ligand; however this will decrease the ratio of bound-to-free ligand.

17. Polarization, expressed in "milliP" units that equal the polarization multiplied by 1,000, is generally used by multi-well plate readers as output. The advantage of anisotropy is that in the case that the sample contains several species with different rotational properties, the measured anisotropy is a weighted sum of the anisotropies of these species. In the case of polarization the equations are rather complex.

Acknowledgements

We thank the Swiss National Science Foundation (grant 31003A-118148) for financial support, GIMB for the gift of GR-119566X and GR-186741X, Dr. A. Surprenant (GIMB) for electrophysiological measurements, and all members of the Laboratory of Physical Chemistry of Polymers and Membranes (EPFL) that contributed to the work presented.

References

1. Baindur N, Triggle DJ (1994) Concepts and progress in the development and utilization of receptor-specific fluorescent ligands. Med Res Rev 14:591–664

2. Baindur N, Triggle DJ (1994) Selective fluorescent ligands for pharmacological receptors. Drug Dev Res 33:373–398

3. Kuder K, Kiéc-Kononowicz K (2008) Fluorescent GPCR ligands as new tools in pharmacology. Curr Med Chem 15:2132–2143

4. McGrath JC, Arribas S, Daly CJ (1996) Fluorescent ligands for the study of receptors. Trends Pharmacol Sci 17:393–399

5. Middleton RJ, Kellam B (2005) Fluorophore-tagged GPCR ligands. Curr Opin Chem Biol 9:517–525

6. Colquhoun D (1998) Binding, gating, affinity and efficacy: the interpretation of structure-activity relationships for agonists and of the effects of mutating receptors. Br J Pharmacol 125:924–947

7. Vernekar SKV, Hallaq HY, Clarkson G, Thompson AJ, Silvreti L, Lummis SC, Lochner M (2010) Toward biophysical probes for the 5-HT3 receptor: structure-activity relationship study of granisetron derivatives. J Med Chem 53:2324–2328

8. Lummis SC, Martin IL (1992) Solubilization, purification, and functional reconstitution of 5-hydroxytryptamine3 receptors from N1E-115 neuroblastoma cells. Mol Pharmacol 41:18–23

9. Boess FG, Lummis SC, Martin IL (1992) Molecular properties of 5-hydroxytryptamine3 receptor-type binding sites purified from NG108-15 cells. J Neurochem 59:1692–1701

10. Tairi AP, Hovius R, Pick H, Blasey H, Bernard A, Surprenant A, Lundstrom K, Vogel H (1998) Ligand binding to the serotonin 5-HT3 receptor studied with a novel fluorescent ligand. Biochemistry 37:15850–15864

11. Wohland T, Friedrich K, Hovius R, Vogel H (1999) Study of ligand-receptor interactions by fluorescence correlation spectroscopy with different fluorophores: evidence that the homopentameric 5-hydroxytryptamine type 3As receptor binds only one ligand. Biochemistry 38:8671–8681

12. Langmuir I (1918) The absorption of gases on plane surfaces of glass, mica and platinum. J Am Chem Soc 40:1361–1403

13. Cheng Y, Prusoff WH (1973) Relationship between the inhibition constant (K1) and the concentration of inhibitor which causes 50 per cent inhibition (I50) of an enzymatic reaction. Biochem Pharmacol 22:3099–3108

14. Briddon SJ, Hill SJ (2007) Pharmacology under the microscope: the use of fluorescence correlation spectroscopy to determine the properties of ligand-receptor complexes. Trends Pharmacol Sci 28:637–645

Chapter 13

Imaging Single Synaptic Vesicles in Mammalian Central Synapses with Quantum Dots

Qi Zhang

Abstract

This protocol describes a sensitive and rigorous method to monitor the movement and turnover of single synaptic vesicles in live presynaptic terminals of mammalian central nervous system. This technique makes use of Photoluminescent semiconductor nanocrystals, quantum dots (Qdots), by their nanometer size, superior photoproperties, and pH-sensitivity. In comparison with previous fluorescent probes like styryl dyes and pH-sensitive fluorescent proteins, Qdots offer strict loading ratio, multi-modality detection, single vesicle precision, and most importantly distinctive signals for different modes of vesicle fusion. Qdots are spectrally compatible with existing fluorescent probes for synaptic vesicles and thus allow multichannel imaging. With easy modification, this technique can be applied to other types of synapses and cells.

Key words Quantum dot, Synaptic vesicle, Full-collapse fusion, Kiss-and-run, Neurotransmitter

1 Introduction

Chemical synapses are specialized cell connections conducting over 95% of rapid communication between excitable cells. They are composed of presynaptic terminals which contain hundreds of synaptic vesicles, tiny membrane-bound organelles filled with neurotransmitter molecules at high concentration (1, 2), and the postsynaptic density, which contains receptors for those neurotransmitters. Whether the communication is neuron to neuron or neuron to muscle, excitatory, inhibitory, or modulatory, fundamental aspects of the unitary signaling event appear to be similar and highly refined. The sequence of events that underlies quantal transmission can be illustrated for central excitatory synapses, typical of those that support sensation, action, learning, and memory in the brain (Fig. 1) (3). The early steps include the arrival of membrane depolarization at the presynaptic structure (nerve terminal or synaptic bouton) and the opening of voltage-gated Ca^{2+} channels. Ca^{2+} influx through these channels leads to a surge of Ca^{2+} in nanodomains

Matthew R. Banghart (ed.), *Chemical Neurobiology: Methods and Protocols*, Methods in Molecular Biology, vol. 995, DOI 10.1007/978-1-62703-345-9_13, © Springer Science+Business Media New York 2013

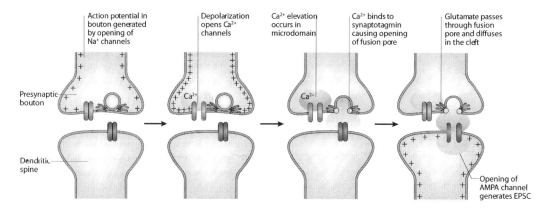

Nature Reviews | Neuroscience

Fig. 1 Steps in the process of chemical synaptic transmission, showing the basic agreement between millisecond events in neurotransmission and their molecular basis. Similar events occur at various synapses in both vertebrates and invertebrates but are illustrated for the case of a glutamatergic synapse of the mammalian CNS. From (3) with permission

near the presynaptic release machinery (4), which includes SNARE proteins and the Ca^{2+} sensors like synaptotagmin I (5). A conformational change in the Ca^{2+} sensors causes fusion between the vesicular and plasma membranes, the so-called exocytosis, which opens the vesicular lumen to the extracellular space. The diffusional spread of neurotransmitter molecules through this opening finally results in rapid activation of ionotropic receptors residing on the postsynaptic cell membrane, which convert the chemical signal (neurotransmitter) back to a change in membrane potential. Clearly, the central event linking the early and late steps is neurotransmitter release, wherein synaptic vesicles discharge their contents into the narrow cleft between the presynaptic and postsynaptic membranes. This requires the establishment of continuity between the vesicle interior and the cleft via a passageway called the fusion pore (6, 7). The nature of the continuity has important implications for how the contents of the vesicle are released and for the fate of the vesicle membrane, topics of intensive research that are ongoing. The most conventional model of vesicle fusion is the full-collapse fusion (FCF) model, in which the entire contents of the vesicle are released in an irreversible process. A more unconventional model is known as kiss-and-run (K&R), in which the vesicle is thought to release its contents, yet retain enough identity to be reused again and may serve a different function from FCF. However, the significance of K&R and vesicle reuse in mammalian central nerve terminals has remained uncertain despite considerable effort (8–11).

Studies of fusion modes and their possible importance have benefited greatly from fluorescent probes. Styryl dyes like FM1-43 fluoresce when inserted in the hydrophobic membrane but not in aqueous solution and thereby can be used to report vesicle fusion events (12). pHluorin, a pH-sensitive GFP, can be fused to the luminal domain of a vesicular protein like synaptobrevin II, and reports the deacidification and reacidification of the vesicle lumen, results of vesicle fusion and retrieval, respectively (13). These probes focus on particular aspects of fusion events, involving lipids or vesicular proteins, and have supplied valuable insights into vesicle dynamics. However, neither FM dyes nor pHluorin fusion proteins provide a signal that crisply distinguishes K&R from FCF. FM dyes can leave the vesicle membrane by various routes such as aqueous departitioning and escape through the fusion pore during K&R (10, 14), or by lateral diffusion away from fused membrane upon FCF (15). Clusters of pHluorin-fused vesicular proteins can persist in two ways: corralled in the vesicle during K&R, or tethered by auxiliary proteins even after FCF (16). More importantly, these probes are not responsive to vesicle shape, fundamental to the original definitions of FCF and K&R (17–19).

To describe the dynamic properties and functional impact of K&R, the field requires a more direct and reliable method. We sought an approach to discriminate sharply between FCF and K&R, focusing on a fundamental distinction: the degree of opening of the vesicle lumen. Inspired by the finding that chromaffin granules use K&R to release small molecules (catecholamines) and FCF to discharge their dense peptide core (20), we used an artificial cargo of appropriate size: quantum dots (Qdots). Like probes for sizing the pore regions of ion channels, Qdots could gauge the narrowest aperture in the path between vesicle lumen and external medium, escaping only if the vesicle undergoes drastic loss of shape, i.e., FCF. In addition to their superior photoproperties, the emission intensity of Qdots is sensitive to environmental factors such as pH (21). Qdots emitting at 605 nm have a narrow emission spectrum that fits neatly between the emission peaks of EGFP and FM4-64. This allows concurrent visualization of Qdots and the other optical probes. With an organic coating bearing carboxyl groups, the Qdots have a hydrodynamic diameter of 15 nm, determined with quasi-elastic light scattering (22, 23). This size is appropriate in multiple respects (Fig. 2): it is small enough for only one Qdot to fit into a synaptic vesicle, but large enough to be completely rejected by K&R fusion pores (1–3 nm) (24). According to these unique features, single Qdots residing in individual vesicles provided two sharply distinct signals for K&R and FCF (Fig. 2), enabling us to show how the prevalence of K&R and the duration of fusion pore opening are regulated by firing frequency, and how vesicles are re-acidified before undergoing reuse (23). Furthermore, individually Qdot-labeled synaptic vesicles can be followed with nanometer precision and thus yield valuable information about vesicle mobility within and between adjacent synapses prior to and immediately after fusion. Last but not the least, correlated

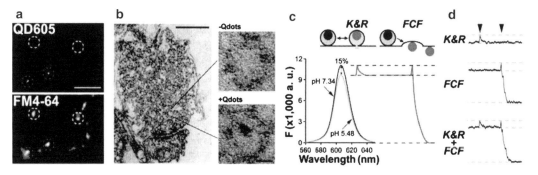

Fig. 2 (**a**) Qdot-labeled single vesicles in presynaptic terminals outlined by subsequent FM4-64 staining. Scale bar, 3 μm. (**b**) Electron micrographs show Qdots loaded into synaptic vesicles. Scale bars, 400 nm (*right*) and 10 nm (*left*). (**c**) Spectrophotometer-measured Qdot fluorescence change at different pHs and hypothesized Qdot signals corresponding to K&R and FCF. (**d**) Sample recordings of field stimulus (*arrowhead*)-evoked changes of Qdot fluorescence

investigation using electron microscopy can reveal the geometric relationship between recycling vesicles and active zones at presynaptic terminals (i.e., labeled with Qdots), an essential indicator for the kinetics of neurotransmitter release.

2 Materials

One of the most exciting developments in the arena of fluorescent probes is the introduction of photoluminescent semiconductor Qdots. Qdots are single particles composed of a crystal core and a shell made of semiconductor. As the core size varies between 2 and 10 nm, their emission increases in parallel, also known as size-tunable emission. For biological applications, Qdots are coated with hydrophilic materials to make them water soluble and to allow conjugation with biomolecules such as proteins and nucleotides. Qdots offer unique physical and optical features that are particularly desirable in single-molecular imaging. First, their nanometer size allows Qdot-based probes to access crowded cellular spaces such as the synaptic cleft, which latex beads cannot enter (25). Second, they fluoresce hundreds of times brighter than organic fluorophores, which translates to an improvement of two orders in magnitude in the precision of single-particle tracking. Third, it is nearly impossible to photobleach Qdots, which thereby permits much longer periods of live imaging. Fourth, multicolor imaging is much easier using Qdots because of their broad absorbance and narrow emission. Fifth, Qdots are more electron dense than biological structures and therefore offer the possibility of correlated optical and electron microscopy imaging for ultrastructural studies.

2.1 Rat Hippocampal Culture

1. Minimum essential medium (MEM) (Invitrogen).
2. D-Glucose.
3. $NaHCO_3$.
4. 200 mM L-glutamine (Invitrogen).
5. Fetal bovine serum (FBS, Invitrogen).
6. Transferrin (Calbiochem).
7. Insulin (Sigma).
8. Culture medium (500 mL MEM, 2.5 g glucose, 100 mg $NaHCO_3$, 50 mg transferrin, 50 mL FBS, 5 mL 0.2 M L-glutamine, 1 mL B27 supplement, 12.5 mg insulin; 0.2 μm filter sterilize and aliquot in 50 mL tubes).
9. Hank's Balanced Salt Solution (HBSS).
10. 0.025% Trypsin (Invitrogen).
11. Cytosine β-D-arabinofuranoside (Ara-C, Sigma).
12. B27 supplement (Invitrogen).
13. Dissection tools.
14. Size 0 circular glass coverslips (Fisher Scientific).
15. Cell culture dishes.
16. Glass pipettes (fire polished before use).
17. Refrigerated centrifuge.
18. Postnatal Day 0–2 Sprague-Dawley rats (Charles River).
19. CO_2 incubator for cell culture.

2.2 Purification of QDots

1. Size-exclusion gel (e.g., Superdex 200, Sigma).
2. Size exclusion column (VWR).
3. Sample concentrator (e.g., Vivaspin 2, Sigma).
4. 8 μM carboxyl Qdot 605 (Invitrogen).
5. 1 M HEPES.
6. 1 M HCl.
7. 1 M NaOH.
8. NaCl.
9. KCl.
10. 1 M $MgCl_2$.
11. 1 M $CaCl_2$.
12. 1,2,3,4-Tetrahydro-6-nitro-2,3-dioxo-benzo(f)quinoxaline-7-sulfonamide hydrate (NBQX, Sigma).
13. 2-amino-5phosphovaleric acid (APV, Sigma).
14. Tyrode solution: 146 mM NaCl, 4 mM KCl, 2 mM $CaCl_2$, 2 mM $MgCl_2$, 10 mM D-glucose, 10 mM HEPES

(310–315 mosm), with pH set at 7.35 with NaOH. 5 μM NBQX and 50 μM D-APV are added to prevent recurrent activity and the development of synaptic plasticity during stimulation.

2.3 Labeling Synaptic Vesicles with Qdots and FM 4-64

1. Bent-tip forceps.

2. Recording chamber that allows optical imaging and electrical field stimulation (e.g., D6RG chamber and PH1 platform, Warner Instrument).

3. Inverted microscope equipped with a high numerical aperture (NA) objective (NA > 1.3), a mercury lamp (100 W) or a xenon lamp (75 W), and appropriate filter sets. We use inverted microscopes from Nikon (Ti-E) and an oil-immersion objective (Nikon X60 NA = 1.49). Excitation and emission filters are as follows: D470/40 excitation and HT605/20 emission for Qdot, D490/20 excitation and D660/50 emission for FM 4-64 (all filters are from Chroma Technology Corp) (see Note 1).

4. Vacuum pump for perfusion.

5. Field stimulation is delivered via a pair of electrodes placed above the submerged cell culture and controlled by a micromanipulator. The electrodes are connected to a stimulation isolator (Warner Instruments). The input channels are connected to a Digidata 200B and controlled by P-clamp software (Molecular Device).

6. Styryl dye FM 4-64: *N*-(3-triethylammoniumpropyl)-4-(6-(4-(diethylamino)phenyl)hexatrienyl)pyridinium dibromide (Invitrogen).

7. 5 mM FM 4-64 stock in DMSO.

8. Imaging software (e.g., Metamorph).

9. Image analysis software (e.g., Image J).

3 Methods

To investigate central synapses where K&R can contribute the most, we use rodent hippocampal neurons 12–21 days in vitro (DIV). This preparation has been used to study the physiology of synaptic vesicles and neurotransmitter release for decades. A large collection of literature and methodologies are readily available. Due to its simple configuration, it is idea for real-time fluorescence imaging with high numeric aperture objectives. A number of Qdot-based imaging studies have been performed using this preparation.

3.1 Rat Hippocampal Culture (as Described in (26) with Little Modification)

1. Dissect hippocampi from postnatal day 0–2 Sprague-Dawley rats and isolate CA1 and CA3 region. Cut into small pieces (1 mm × 1 mm).

2. Wash tissue 3× with HBSS and digest with trypsin.

3. Wash tissue 3× with HBSS and titrate with a fire-polished glass pipette.

4. Centrifuge for 10 min, $800 \times g$, 4°C.

5. Aspirate supernatant and resuspend cell pellet with culture medium.

6. Plate cell on glass coverslips in culture plates or dishes.

7. Add additional culture medium up to 2 mL after 1 h.

8. Add the mitosis inhibitor ARA-C 1–2 days after cell plating to stop glial growth.

9. Change the culture medium twice per week and the cells will be ready after 12–18 DIV.

3.2 Purification of Qdots

1. Centrifuge with maximum speed (e.g., $10,000 \times g$) for 15 min at 4°C. *This step moves most of the Qdot aggregates.

2. Gel filtration of Qdot with size exclusion column (e.g., Superdex 200) and collect all eluted solutions.

3. Concentrate it with Vivaspin 2 mL by adding all eluted solutions in the concentrator chamber and spin at $10,000 \times g$, 4°C for 10 min.

4. Remove collection tube and resuspend Qdots with appropriate volume of Tyrode solution to have the final stock solution of Qdot reconstituted at 8 µM.

3.3 Labeling Synaptic Vesicles with Qdots

1. Use bent-tip forceps to carefully pick up coverslips on which cells are cultured and place them in sealed imaging chamber containing 400 µL Tyrode solution.

2. Place imaging chamber on the microscope stage and connect the gravity-vacuum perfusion tubing.

3. Turn on gravity perfusion and set dripping speed at about 0.2 mL/s.

4. Turn on transmitter light and search for field of imaging. The criterion is to avoid neuronal soma and area clustered with dendrites and axons.

5. Stop perfusion and keep coverslips submerged in ~400 µL Tyrode solution.

6. Dilute Qdot stock solution to 80 nM in 400 µL Tyrode solution for whole bouton labeling or 0.2 nM in 400 µL Tyrode solution for single vesicle loading.

7. Slowly drop the 400 µL Qdot-containing Tyrode solution in the imaging chamber and mix gently.

8. Submerge the stimulation electrodes into Tyrode solution.

9. Deliver a train of field stimulation (e.g., 60-s resting period, 10-Hz 2-min spike train, and 60-s resting period).

10. Switch on perfusion and wash the cells for 15–20 min. To ensure complete removal of unloaded Qdots, periodically examine the Qdot signals using eyepiece. Presence of Qdots outside of dendrite or axon areas is the sign that the wash-off is not complete.

3.4 Labeling Synaptic Vesicles with FM 4-64 (This can be Done Simultaneously with Qdot or Retrospectively)

1. Dilute FM 4-64 stock solution to 5 µM in 400 µL Tyrode solution.

2. Slowly drop the 400 µL FM 4-64-containing Tyrode solution in the imaging chamber and mix gently.

3. Submerge the stimulation electrodes into Tyrode solution.

4. Deliver a train of field stimulation (e.g., 60-s resting period, 10-Hz 2-min spike train, and 60-s resting period).

5. Switch on perfusion and wash the cells for 5–10 min. To ensure complete removal of unloaded FM4-64, periodically examine the FM 4-64 fluorescence. Presence of FM 4-64 outside along cell surface is a sign of incomplete wash-off (see Note 2).

3.5 Imaging

1. During the washing period, adjust imaging settings including exposure time (e.g., 100 ms), frame interval (e.g., 333 ms for 3 Hz imaging rate), and video gain (i.e., electron-multiplication) such that the maximum photoluminescence signal of Qdots reaches 80% of the dynamic range of the detection device (e.g., EMCCD).

2. Set up the synchronization of imaging rate and stimulation trigger. For example, if field electrical stimulation is applied, set the trigger of the field stimulation protocol as the opening of the imaging shutter. In our experiments, we delivered a TTL pulse from the imaging program upon the first opening of the excitation shutter. This pulse serves as an external trigger for the start of a 0.1-Hz 2-min field stimulation train with a 30-s pre-stimulation delay for baseline measurement.

3. Choose regions of interests (ROIs) for fluorescence measurement. This can be done online or off-line. If the imaging programs allow (e.g., MetaFluor), it is better to do it online so that imaging and the first step of data analysis can be conducted simultaneously. We use circular ROIs with fixed positions and the same size cover for all synaptic boutons identified by over-expressed vesicular proteins fused with fluorescent proteins or co-labeled with FM4-64. The size of ROIs is

~2 μm, which is twice the average size of CNS presynaptic terminals. In images, the sizes of the ROIs are measured as the number of pixels. The size of the pixels is determined by magnification and pixel size of the photon detector.

4. Start continuous imaging and store the imaging stack on the computer RAM. Upon finishing, immediately rename the imaging file encoding experiment information and save it on the hard drive. If online ROI measurement is conducted, export the fluorescence intensity data to appropriate file format (e.g., Microsoft Excel) and save it accordingly.

3.6   Data Analysis

1. *Single synaptic vesicle tracking.* The trajectories of individual Qdot-labeled synaptic vesicles are reconstructed from the image stacks. Single vesicles are identified by their Qdot photoluminescence. A variety of open-source software is available for such analysis such as ImageJ (http://rsbweb.nih.gov/ij/) and ImageJ-based Fiji (http://pacific.mpi-cbg.de/wiki/index.php/Fiji). A Qdot-specialized tracking program, SINEMA (http://www.lkb.ens.fr/recherche/optetbio/tools/sinema/), is used in some of our studies. In brief, this is a two-step process that is applied successively to each frame of the image stack. First, bright spots are detected by crosscorrelating the image with a Gaussian model of the point spread function. A least-squares Gaussian fit is employed to determine the center of each spot and the spatial accuracy is determined by the signal-to-noise ratio. Second, Qdot trajectories are assembled automatically by linking, from frame to frame, the computer-determined centers from the same Qdots. If blinking occurs, the association criterion is based on the assumption of free Brownian diffusion and missing frames are linked by straight line for simplification. In addition, a post-analysis manual association step is performed to exclude wrongly associated trajectories from neighboring Qdots.

2. *Distinguishing single FCF and K&R events.* This is a three-step procedure including digitizing the fluorescence signal, identifying stimulation-evoked fusion events and spontaneous events, and distinguishing FCF and K&R based on the pattern of fluorescence change. First, average fluorescence intensities in the ROIs are extrapolated and stored as absolute numbers in a data file. Background fluorescence is measured and averaged from four non-Qdot areas of each frame. The Qdot photoluminescence change thus can be extracted by subtracting background value from absolute values. The pre-stimulation and post-stimulation baseline can be determined by averaging Qdot fluorescence before and after field stimulation. Second, the Qdot signal is aligned with field stimulation pulses along the timeline. All events that exhibit a 15% Qdot photoluminescence

jump within 300 ms (i.e., one imaging frame with 3-Hz imaging rate) after the field stimulation pulse is designated as evoked fusion events and all other events are considered spontaneous events. Third, FCF and K&R events are manually identified, i.e., transient increase of 15% unitary Qdot photoluminescence as K&R and 15% transient increase followed by unitary loss of Qdot photoluminescence as FCF. All fusion events of every Qdot labeled can be further digitized for a raster plot or further analysis.

4   Anticipated Results

Figure 2 shows an example of cultured hippocampal neurons labeled with Qdots. Presynaptic boutons are identified by retrospective staining with FM 4-64. The signals of FM 4-64 and Qdots can be separated using appropriate filter sets. The emission of Qdots and other fluorescence dyes or proteins can also be detected simultaneously using optical devices like Dual-View (Photometrics).

Figure 3 shows sample traces representing single K&R and FCF evoked by field stimulation. In addition, hundreds of events from Qdot-labeled single synaptic vesicles are summarized in a digitized raster plot to obtain an overview of the population behavior of synaptic vesicle fusion at mammalian central nervous system synapses. Increasing the imaging rate to 30-Hz reveals further details of fusion kinetics such as the duration of fusion pore open and reacidification of retrieved vesicles.

5   Additional Considerations

The current protocol has two major limitations. First, the mechanism of vesicular uptake of Qdots, especially the parameters that determine Qdots' affinity for neuronal membranes, is not completely understood. We know that either a negative charge or a carboxyl group itself is necessary for such affinity. But we do not know the target of the attachment and thus lack an effective way to control this membrane affinity. However, by systematic adjustment of coating properties, it is possible to optimize this key parameter such that Qdots can bind to neuronal membrane, but not too tightly, so that they can readily depart from the membrane after exiting vesicles. The second limitation results from the intrinsic photo-intermittency of Qdots, a.k.a. "blinking." Although blinking events provide an independent criterion to quantify unitary Qdot photoluminescence, the randomness of blinking duration and interval obscure measurement. To solve this problem, specific algorithms for overcoming blinking events are rapidly evolving (27). With the continuous development of Qdots and the advancement

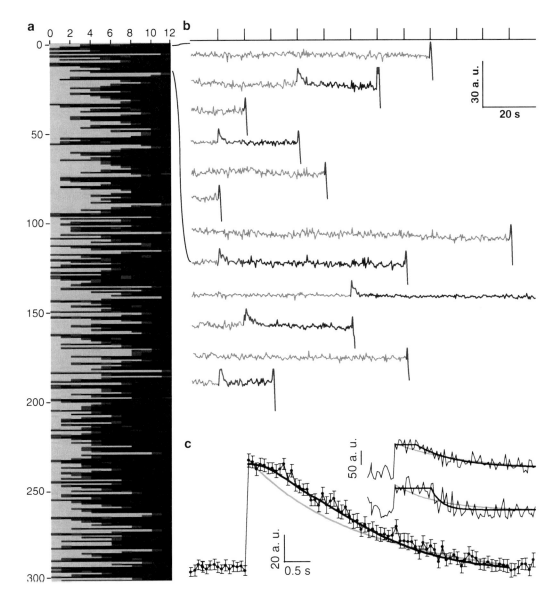

Fig. 3 (**a**) Raster representation of traces ($n = 302$) from single Qdot-loaded vesicles that responded to 0.1-Hz field stimulation for 2 min. For each stimulus and subsequent interval, Qdot signals registered as nonresponse (*gray*), K&R (*red*), nonresponse after K&R (*maroon*), FCF (*blue*), or Qdot no longer present in the region of interest (*black*). Pooled traces from $N = 8$ coverslips, three separate cultures. (**b**) Traces corresponding to the first 12 rasters in (**a**). Photoluminescence changes are color coded for each stimulus and ensuing interval as in (**a**). (**c**) Comparison of pooled data exemplified by insets (*black symbols* with error bars indicating SEM, $n = 43$) and averages of the two kinds of fits (*gray* and *black*). *Insets*: Samples taken in normal Tyrode solution with single shocks; interstimulus interval, >20 s. Two types of fits were overlaid: a single-exponential decay (*gray*) and a plateau followed by an exponential (*black*), the latter fitting significantly better even after statistical penalization for the extra parameter (Akaike information criterion score, −60.5; $P < 0.001$)

of super-resolution imaging techniques, it is certain that Qdot-based single-molecule imaging will become a breakthrough in cellular neuroscience.

6   Notes

1. QDs can be excited at any wavelength shorter than 550 nm. Use a wide band-pass excitation filter to collect maximum light power for excitation. Avoid excitation with ultraviolet (UV) light, which leads to photodamage to the cells and increased blinking of QDs. A narrow-band emission filter avoids cross talk of signals from different fluorophores. Sensitive image acquisition system, such as an EMCCD (e.g., iXon+897 from Andor).

2. FM4-64 emission is close to far-red and thus hard to see by eye. One can use live imaging mode to monitor the loading and wash-off of FM4-64. However, prolonged exposure may significantly reduce the emission of FM4-64 as it is very prone to photo-bleaching.

Acknowledgments

I thank my postdoctoral mentor, Dr. R. W. Tsien, for giving me the opportunity to develop this Qdot-based vesicle labeling method in his lab. I also thank Invitrogen for providing the documentation of Qdots' properties. This work was supported by grants from NIDA and AFAR to Q.Z.

References

1. Katz B (1969) The release of neural transmitter substances. Charles C Thomas, Springfield, IL
2. Suszkiw JB, Zimmermann H, Whittaker VP (1978) Vesicular storage and release of acetylcholine in Torpedo electroplaque synapses. J Neurochem 30(6):1269–1280
3. Lisman JE, Raghavachari S, Tsien RW (2007) The sequence of events that underlie quantal transmission at central glutamatergic synapses. Nat Rev Neurosci 8(8):597–609
4. Llinas R, Sugimori M, Silver RB (1992) Microdomains of high calcium concentration in a presynaptic terminal. Science 256(5057):677–679
5. Sudhof TC (2004) The synaptic vesicle cycle. Annu Rev Neurosci 27:509–547
6. Jackson MB, Chapman ER (2008) The fusion pores of Ca2+ -triggered exocytosis. Nat Struct Mol Biol 15(7):684–689
7. Spruce AE et al (1990) Properties of the fusion pore that forms during exocytosis of a mast cell secretory vesicle. Neuron 4(5):643–654
8. Aravanis AM, Pyle JL, Tsien RW (2003) Single synaptic vesicles fusing transiently and successively without loss of identity. Nature 423(6940):643–647
9. He L et al (2006) Two modes of fusion pore opening revealed by cell-attached recordings at a synapse. Nature 444(7115):102–105
10. Richards DA, Bai J, Chapman ER (2005) Two modes of exocytosis at hippocampal synapses revealed by rate of FM1-43 efflux from individual vesicles. J Cell Biol 168(6):929–939
11. Serulle Y, Sugimori M, Llinas RR (2007) Imaging synaptosomal calcium concentration microdomains and vesicle fusion by using total internal reflection fluorescence microscopy. Proc Natl Acad Sci U S A 104(5):1697–1702

12. Betz WJ, Mao F, Bewick GS (1992) Activity-dependent fluorescent staining and destaining of living vertebrate motor nerve terminals. J Neurosci 12:363–375

13. Miesenböck G, De Angelis DA, Rothman JE (1998) Visualizing secretion and synaptic transmission with pH-sensitive green fluorescent proteins. Nature 394(6689):192–195

14. Aravanis AM et al (2003) Imaging single synaptic vesicles undergoing repeated fusion events: kissing, running, and kissing again. Neuropharmacology 45(6):797–813

15. Zenisek D et al (2002) A membrane marker leaves synaptic vesicles in milliseconds after exocytosis in retinal bipolar cells. Neuron 35(6):1085–1097

16. Willig KI et al (2006) STED microscopy reveals that synaptotagmin remains clustered after synaptic vesicle exocytosis. Nature 440(7086): 935–939

17. Ceccarelli B, Hurlbut WP, Mauro A (1973) Turnover of transmitter and synaptic vesicles at the frog neuromuscular junction. J Cell Biol 57(2):499–524

18. Fesce R et al (1994) Neurotransmitter release: fusion or 'kiss-and-run'? Trends Cell Biol 4(1):1–4

19. Heuser JE, Reese TS (1973) Evidence for recycling of synaptic vesicle membrane during transmitter release at the frog neuromuscular junction. J Cell Biol 57(2):315–344

20. Fulop T, Radabaugh S, Smith C (2005) Activity-dependent differential transmitter release in mouse adrenal chromaffin cells. J Neurosci 25(32):7324–7332

21. Gao X, Chan WC, Nie S (2002) Quantum-dot nanocrystals for ultrasensitive biological labeling and multicolor optical encoding. J Biomed Opt 7(4):532–537

22. Zhang Q, Cao YQ, Tsien RW (2007) Quantum dots provide an optical signal specific to full collapse fusion of synaptic vesicles. Proc Natl Acad Sci U S A 104(45):17843–17848

23. Zhang Q, Li Y, Tsien RW (2009) The dynamic control of kiss-and-run and vesicular reuse probed with single nanoparticles. Science 323(5920):1448–1453

24. Jackson MB, Chapman ER (2006) Fusion pores and fusion machines in Ca^{2+}-triggered exocytosis. Annu Rev Biophys Biomol Struct 35:135–160

25. Dahan M et al (2003) Diffusion dynamics of glycine receptors revealed by single-quantum dot tracking. Science 302(5644):442–445

26. Liu G, Tsien RW (1995) Synaptic transmission at single visualized hippocampal boutons. Neuropharmacology 34(11):1407–1421

27. Bonneau S, Dahan M, Cohen LD (2005) Single quantum dot tracking based on perceptual grouping using minimal paths in a spatiotemporal volume. IEEE Trans Image Process 14(9):1384–1395

Chapter 14

Directed Evolution of Protein-Based Neurotransmitter Sensors for MRI

Philip A. Romero, Mikhail G. Shapiro, Frances H. Arnold, and Alan Jasanoff

Abstract

The production of contrast agents sensitive to neuronal signaling events is a rate-limiting step in the development of molecular-level functional magnetic resonance imaging (molecular fMRI) approaches for studying the brain. High-throughput generation and evaluation of potential probes are possible using techniques for macromolecular engineering of protein-based contrast agents. In an initial exploration of this strategy, we used the method of directed evolution to identify mutants of a bacterial heme protein that allowed detection of the neurotransmitter dopamine in vitro and in living animals. The directed evolution method involves successive cycles of mutagenesis and screening that could be generalized to produce contrast agents sensitive to a variety of molecular targets in the nervous system.

Key words Magnetic resonance imaging, Directed evolution, Protein engineering, Cytochrome P450, Dopamine

1 Introduction

With the continuing development of magnetic resonance imaging (MRI) contrast agents sensitive to molecular hallmarks of brain activity (1), a new generation of functional neuroimaging methods could combine noninvasive, whole-brain coverage with unprecedented specificity for neuronal processes. Design and synthesis of effective contrast agents for molecular fMRI remains a significant challenge, however. Protein engineering strategies applied to MRI-detectable proteins (2–5) might provide a means to accelerate this endeavor. Many proteins contain paramagnetic ions that can enable them to be visualized by MRI with longitudinal (T_1) or transverse (T_2) relaxation time weighting. Advantages of protein contrast agents, compared with conventional synthetic agents, include amenability to modification, ease and low cost of production, and the

---

Philip A. Romero and Mikhail G. Shapiro contributed equally to this work.

Matthew R. Banghart (ed.), *Chemical Neurobiology: Methods and Protocols*, Methods in Molecular Biology, vol. 995, DOI 10.1007/978-1-62703-345-9_14, © Springer Science+Business Media New York 2013

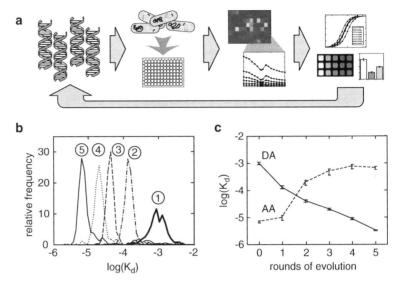

Fig. 1 Overview of the directed evolution strategy. (**a**) The parental protein gene is randomly mutated and the resulting variants are transformed into *E. coli*. This library is grown and expressed in 96-well plates. High-throughput neurotransmitter binding assays are then performed in cell lysate. Several of the top variants identified in the high-throughput binding assay are purified and characterized in greater detail. The best variant from each generation is used as the parent in the next generation, and this process is repeated until the desired properties are achieved. (**b**) Distribution of apparent dissociation constants (K_d) for dopamine measured from mutant BM3h proteins generated over five rounds of directed evolution for enhanced dopamine affinity and decreased affinity for the native ligand, arachidonic acid (data from ref. 6). Numbers in circles denote the round of evolution associated with each mutant library. (**c**) K_d values for dopamine (DA, *solid line*) and arachidonic acid (AA, *dashed line*), for wild-type BM3h and mutant variants isolated at each round of screening as in (**b**)

possibility of targeting and endogenous production in genetically modified cells or organisms.

In recent work, we used the protein engineering technique of directed evolution to generate dopamine-sensing mutants of the cytochrome P450 BM3 heme domain (BM3h) (6). Our approach involved applying repeated cycles of random mutagenesis and screening, each step allowing us to identify BM3h mutants with enhanced affinity and selectivity for dopamine, compared with other potential ligands (Fig. 1). After four rounds of successive improvements, we obtained a variant with an 8.9 μM dissociation constant for dopamine and little responsiveness to other neurotransmitters. This molecule acted as an MRI contrast agent with T_1 relaxivity (r_1) values of 1.1 ± 0.1 and 0.17 ± 0.03 mM^{-1}s^{-1} in the absence and presence of saturating dopamine concentrations, respectively. The agent responded semiquantitatively to dopamine release from cultured PC12 cells in vitro, and allowed detection of dopamine dynamics in the brains of living rodents.

The directed evolution method is highly versatile (7), and could be used to prepare MRI contrast agents based on BM3h or other paramagnetic proteins; these probes could be selected for sensitivity to a wide variety of neurochemicals. Analytes of interest might include neurotransmitters other than dopamine, cytokines, or components of intracellular signal transduction pathways. Here we detail the protocol we applied to produce BM3h-based dopamine sensors (6), with possible variations noted where appropriate.

2 Materials

2.1 Preparation of Random Mutagenesis Library

2.1.1 Library Construction

1. Plasmid containing parent gene for directed evolution (see Note 1).

2. Appropriate PCR primers for amplification of parent gene.

3. 20× dNTP stock (4 mM dATP, 4 mM dGTP, 10 mM dTTP, 10 mM dCTP).

4. *Taq* DNA polymerase with supplied 10× PCR buffer II and 25 mM $MgCl_2$ (AmpliTaq, Applied Biosystems, Foster City, CA).

5. 500 μM $MnCl_2$ in water (sterilized before use).

6. PCR purification kit (QIAquick PCR Purification Kit, QIAGEN, Valencia, CA).

7. Appropriate restriction enzymes with supplied buffers (New England Biolabs, Ipswich, MA).

8. 1% agarose gel containing 0.5 μg/mL ethidium bromide.

9. Gel extraction kit (QIAquick Gel Extraction Kit, QIAGEN, Valencia, CA).

10. Plasmid backbone (digested, gel-purified, and ready for insert ligation).

11. T4 DNA ligase with supplied buffers (New England Biolabs, Ipswich, MA).

12. Electrocompetent BL21(DE3) *E. coli* (E. cloni EXPRESS Competent Cells, Lucigen Corporation, Middleton, WI).

13. LB agar plates containing 100 μg/mL ampicillin.

2.1.2 Library Expression

1. LB medium containing 100 μg/mL ampicillin.

2. Sterile toothpicks.

3. Sterile 96 deep-well plates (2 mL volume).

4. Sterile 50% glycerol.

5. Sterile 96-well microtiter plates.

6. TB medium containing 100 μg/mL ampicillin and 200 μM IPTG.

2.2 High-Throughput Screening

2.2.1 Lysate Preparation

1. Lysis buffer: Phosphate buffered saline (PBS) containing 0.75 mg/mL hen egg lysozyme and 5 mg/mL DNAse I (Sigma-Aldrich, St. Louis, MO).

2. 96-well microtiter plates.

3. PBS.

4. Neurotransmitters of interest (see Note 2).

2.3 Bulk Protein Expression and Purification

2.3.1 Protein Expression

1. LB medium containing 100 µg/mL ampicillin.

2. Sterile culture tubes.

3. TB medium containing 100 µg/mL ampicillin.

4. Sterile culture flask.

5. 1,000× IPTG solution (500 mM in water, filter sterilized).

2.3.2 Protein Purification

1. PBS.

2. Protein extraction kit (BugBuster Plus Lysonase Kit, EMD Chemicals, San Diego, CA).

3. Ni-NTA agarose (QIAGEN, Valencia, CA).

4. Disposable polypropylene column.

5. Ni-NTA wash buffer: 20 mM Tris, 100 mM NaCl, 20 mM imidazole, pH 8.

6. Ni-NTA elution buffer: 20 mM Tris, 100 mM NaCl, 300 mM imidazole, pH 8.

7. PD-10 desalting columns (GE Healthcare, Piscataway, NJ).

8. Amicon Ultra-30 centrifugal filter units (Millipore, Billerica, MA).

2.3.3 Measurement of Protein Concentration

1. Sodium hydrosulfite (Sigma-Aldrich, St. Louis, MO).

2. Carbon monoxide tank (CO is a colorless and odorless gas which is highly toxic).

2.4 Variant Characterization

2.4.1 Neurotransmitter Titrations

1. Neurotransmitters of interest (see Note 2).

2. Cuvettes.

2.4.2 Measurement of MRI Relaxivity

1. Microtiter plates, modified if necessary to fit in an MRI scanner.

2. Neurotransmitters of interest (see Note 2).

3. An MRI scanner (see Note 3).

3   Methods

3.1   Preparation of Random Mutagenesis Library

3.1.1   Library Construction

1. Prepare error-prone PCR by combining the following in a thin-walled PCR tube:

   40 μL water.

   10 μL 10× PCR buffer II.

   5 μL 500 μM $MnCl_2$.

   28 μL 25 mM $MgCl_2$.

   1 μL parent BM3h plasmid DNA.

   5 μL of each primer (forward and reverse).

   5 μL of 20× dNTP mix.

   1 μL Amplitaq DNA polymerase.

   For a 100 μL total reaction volume (see Note 4).

2. Perform PCR with the following program:

   (a) 120-s initial melting at 95°C.

   (b) 30-s melting at 95°C.

   (c) 30-s annealing at 57°C.

   (d) 90-s extension at 72°C.

   (e) Repeat steps b–d for 14 cycles.

3. Purify the PCR product using the PCR purification kit.

4. Perform a double digest on the PCR product for 2 h at 37°C with the appropriate restriction enzymes.

5. Run the restriction digest product on a 1% agarose gel, and excise the appropriate DNA fragment. Use the gel extraction kit to separate the DNA from the agarose.

6. Ligate the digested fragment into the cut plasmid backbone using T4 DNA ligase. Allow the ligation reaction to run at 16°C for at least 12 h.

7. Transform the ligation product into electrocompetent BL21(DE3) cells using a 0.1 cm electroporation cuvette. After electroporation, quickly add 1 mL of recovery media to the cells and incubate at 37°C for 1 h.

8. Plate the transformed cells on LB agar plates and incubate at 37°C overnight (see Note 5).

3.1.2   Library Expression

1. Fill sterile 96 deep-well plates with 400 μL of LB containing 100 μg/mL ampicillin per well.

2. Using sterile toothpicks, pick individual *E. coli* colonies from the plated transformation into each well of the 96 deep-well plates. Be sure to inoculate a few wells with the parent (as a positive control) and keep a few wells empty (as a negative control and to detect contamination).

3. Grow 96-well starter cultures shaking at 300 RPM and 37°C overnight.

4. Fill sterile 96 deep-well plates with 1.2 mL of TB containing 100 µg/mL ampicillin and 200 µM IPTG per well. Using a pipetting robot or multichannel pipette, transfer 100 µL/well of each LB starter culture to each TB plate.

5. Fill sterile microtiter plates with 50 µL of 50% glycerol per well. Using a pipetting robot or multichannel pipettor, transfer 50 µL/well of each LB starter culture to each microtiter plate. Store glycerol stocks at –80°C for later use.

6. Grow 96-well TB cultures shaking at 300 RPM and 30°C overnight.

7. Pellet TB cultures by centrifugation at 3,500 RCF for 15 min. Pour off supernatant and freeze cell pellets.

3.2 High-Throughput Screening

3.2.1 Lysate Preparation

1. Thaw frozen cell pellets containing expressed BM3h variants (or controls) at room temperature, add 650 µL/well of lysis buffer, and resuspend the pellets using a pipetting robot (see Note 6).

2. Incubate the lysate at 37°C for 30 min, again resuspend, and then incubate the lysate at 37°C for another 30 min.

3. Clarify the lysate by centrifugation at 5,500 RCF for 15 min.

4. Using a pipetting robot, transfer 200 µL/well clarified lysate to a microtiter plate. The assay requires one microtiter plate per neurotransmitter screened.

3.2.2 Neurotransmitter Titrations

1. Prepare appropriate neurotransmitter serial dilutions in PBS (see Note 7).

2. Read the absorbance of the lysate from 350 to 500 nm (see Note 8).

3. Add 10 µL/well of the lowest concentration neurotransmitter and read absorbance from 350 to 500 nm. Continue to add increasing amounts of neurotransmitter, scanning after each addition. Characteristic spectroscopic results are presented in Fig. 2.

4. Repeat titrations for all desired neurotransmitters.

3.2.3 Data Analysis

1. The data set consists of optical density (OD) readings from 350 to 500 nm for each BM3h variant in the presence of varying concentrations of each neurotransmitter (see Note 9). Binding curves for each variant can be generated by subtracting the baseline absorbance obtained in the absence of neurotransmitter from each spectrum, and then subtracting the minimum from the maximum of each difference spectrum to obtain $\Delta OD_{max–min}$ values (Fig. 2c).

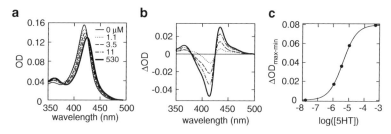

Fig. 2 Spectroscopic BM3h binding assay. (**a**) Absorbance spectra (optical density, OD, vs. wavelength) measured for a BM3h variant in the presence of varying concentrations of a neurotransmitter ligand, in this case serotonin. The spectral change is associated with perturbations to the electronic structure of the BM3h heme, in this case probably resulting from direct coordination of the heme iron by serotonin (12). (**b**) Difference spectra (ΔOD vs. wavelength) obtained by subtracting the ligand-free spectrum in panel (**a**) from spectra measured at each ligand concentration. (**c**) Maximum and minimum values in the difference spectra are subtracted from one another to yield $\Delta OD_{max-min}$ values for each serotonin concentration ([5HT]). K_d estimates are obtained by fitting a binding model to these data

2. Dissociation constant (K_d) estimates for binding of neurotransmitters to each BM3h variant are obtained by fitting the absorbance data to a ligand-depleting bimolecular association function of the form

$$\Delta OD_{max\text{-}min} = k\left[x + p + K_d - \sqrt{(x + p + K_d)^2 - 4xp}\right],$$

where k is a constant of proportionality and x and p are the total analyte and BM3h protein concentrations, respectively (see Note 10).

3. The best 5–10 variants should be chosen for bulk expression and further characterization (see Note 11).

3.3 Bulk Protein Expression and Purification

3.3.1 Protein Expression

1. Using the 96-well glycerol stocks made during step 4 of Subheading 3.1.2, inoculate 5 mL LB starter cultures containing 100 μg/mL ampicillin with the top variants chosen from the high-throughput screen. Grow LB starter cultures shaking at 37°C overnight.

2. Inoculate 50 mL TB cultures with 500 μL of the starter culture and incubate shaking at 37°C. Once the culture reaches an OD_{600} of 0.6, induce with 50 μL of 1,000× IPTG. Reduce the growth temperature to 30°C and allow proteins to express for 20–30 h.

3. After expression, pellet cells by centrifuging at 4,000 RCF for 15 min and then freeze cell pellets.

3.3.2 Protein Purification

1. Resuspend frozen pellet with protein extraction reagents, 5 mL BugBuster, and 10 μL lysonase per gram of protein pellet. Shake for 15–20 min at room temperature (see Note 12).

2. Clarify lysate by centrifuging at 16,000 RCF for 30 min at 4°C.

3. Load clarified lysate onto a column packed with 5 mL Ni-NTA agarose resin.

4. Wash column with 40 mL of Ni-NTA wash buffer.

5. Elute histidine-tagged BM3h variants with 5 mL of Ni-NTA elution buffer.

6. Run Ni-NTA column elution through a PD-10 column to remove imidazole and to exchange buffer to PBS.

7. Concentrate as needed using Amicon Ultra-30 centrifugal filters.

3.3.3 Measurement of Protein Concentration

1. Dilute BM3h samples to about 10 μM in 100 mM Tris buffer, pH 8. Transfer to a cuvette and add a few milligrams of sodium hydrosulfite.

2. Read the absorbance spectra from 400 to 500 nm.

3. Slowly bubble the BM3h solution with CO for 30 s.

4. Read the absorbance spectra from 400 to 500 nm.

5. Calculate the CO-difference spectra by subtracting the pre-CO spectra from the post-CO spectra, and determine the difference between ΔOD values obtained at 450 and 490 nm (Fig. 3). The BM3h concentration can be determined using Beer's Law: $\Delta OD_{450-490} = \varepsilon_{450-490}[BM3h]l$, with the extinction coefficient $\varepsilon_{450-490} = 91$ mM^{-1} cm^{-1} (8), where l is the light path length.

3.4 Characterization of Purified P450 Variants

3.4.1 Neurotransmitter Titrations

1. Prepare concentrated neurotransmitter solutions (see Note 13).

2. Prepare a 1 μM BM3h solution using PBS as a diluent.

3. Add the 1 μM protein to a cuvette and read the absorbance from 350 to 500 nm.

4. Add logarithmically increasing amounts of neurotransmitter and read absorbance from 350 to 500 nm after each addition.

5. Analyze binding data as described in Subheading 3.2.3.

3.4.2 Measurement of Nuclear Magnetic Relaxivity

1. Prepare six to eight BM3h samples with concentrations ranging from 0 to 240 μM using PBS as a diluent.

2. Divide these samples into two, and add saturating amounts of the neurotransmitter of interest to half. Add equivalent amount of buffer to the other half.

3. Fill central wells of a 384-well microtiter plate with 100 μL of each sample (all concentrations with/without neurotransmitter) (see Note 14).

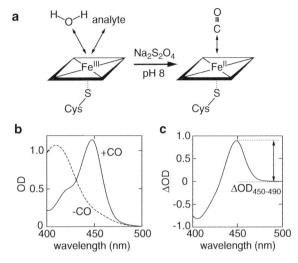

Fig. 3 Quantification of BM3h concentration. (**a**) Binding of analytes near the ferric heme of stably oxidized BM3h (Cys400 ligand shown) modulates relaxivity probably by competing with water molecules (*left*), but also complicates quantification of BM3h concentration by affecting the protein's absorbance spectrum. Robust, analyte-independent concentration measurement is possible by reducing the protein with sodium hydrosulfite and measuring spectra in the presence and absence of carbon monoxide (CO, *right*). (**b**) Sample BM3h spectra obtained before (−CO, *dashed line*) and after (+CO, *solid line*) CO binding. (**c**) The difference spectrum obtained by subtracting the −CO spectrum from the +CO spectrum. Quantification of BM3h concentration is performed by measuring the difference between ΔOD values recorded at 450 and 490 nm ($\Delta OD_{450-490}$), and then applying Beer's law with an extinction coefficient of 91 mM^{-1} cm^{-1} (8)

4. Fill unused wells within two or more wells of the samples with 100 µL PBS.

5. Image the microtiter plates at 21°C using a spin echo pulse sequence with short echo time (TE) and a range of repetition times (TR; see Note 15). Figure 4a shows a sample image.

6. Calculate longitudinal relaxation rates ($1/T_1$) by curve fitting to the reconstructed image data, using an equation of the form $I = k(1 - \exp(-TR/T_1))$, where I is the observed MRI signal intensity and k is a constant of proportionality (Fig. 4b). Values of r_1 are subsequently determined by linear fitting to a plot of $1/T_1$ against protein concentration (Fig. 4c; see Note 16).

3.5 Directed Evolution

Bulk measurements of ligand binding affinity and relaxivity are used to select BM3h variants with the most desirable properties. The above procedures (Subheadings 3.1–3.4) are then repeated over multiple iterations of improvement, until the desired affinity and specificity are achieved (cf. Fig. 1). In each generation 400–800 variants are screened, and within 5–10 generations K_d values might be expected to change by 2–3 orders of magnitude (see Note 17).

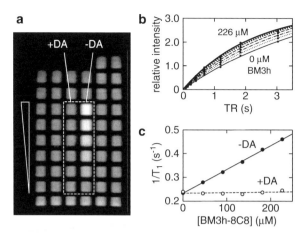

Fig. 4 Measurement of relaxivity by MRI. (**a**) Wells of microtiter plate are filled with protein solutions (*inside dashed rectangle*) or buffer (*outside*). The buffer wells limit magnetic susceptibility-related artifacts in images; extra buffer wells (*top right*) are used to identify the plate orientation. A T_1-weighted spin echo image obtained at 4.7 T is shown here (TE/TR = 10/477 ms), containing two columns of BM3h solutions ranging from 0 to 256 μM (increasing [BM3h] indicated by *wedge*), in the presence (+DA, *left column*) or the absence (−DA, *right column*) of 1 mM dopamine. The BM3h variant used here (BM3h-8C8) was selected for dopamine binding, and intensity differences between the +DA and −DA conditions are readily apparent. (**b**) Quantification of MRI signal in images obtained at multiple TR values enables estimation of the relaxation rate ($1/T_1$) for each condition. An exponential decay curve with time constant T_1 (*lines*) is fitted to each plot of relative MRI intensity vs. TR (*circles*); superimposed plots with fitted curves are shown here for 0–266 μM BM3h-8C8 in the absence of dopamine (cf. *right-hand column* in panel **a**). (**c**) Values of $1/T_1$ in the presence (*open circles*) or the absence (*filled circles*) of dopamine are plotted as a function of protein concentration ([BM3h-8C8]). Longitudinal relaxivity (r_1) values are given by the slopes of linear fits to the data

4   Notes

1. Choosing a good parent gene is one of the most important steps when engineering proteins by directed evolution. The parent protein should be characterized with the high-throughput screen (Subheading 3.2) and bulk assays (Subheadings 3.3 and 3.4). The parent should also display at least marginal levels of the desired activity, in order to allow the assay to resolve small increases in binding affinity.

2. The BM3h spectroscopic binding assay has been performed on numerous neurotransmitters and other small-molecule ligands (6). Neurotransmitters should be chosen with consideration for the neurotransmitter of interest in addition to other small molecules which may competitively bind. Variants are selected for affinity and specificity toward the neurotransmitter of interest.

3. This protocol was developed on a 40-cm-bore Bruker Avance 4.7 T MRI scanner, equipped with 12-cm-inner-diameter 26 G cm^{-1} triple-axis gradients and a 10-cm-inner-diameter birdcage resonator radiofrequency coil. A wide variety of MRI systems could be used for relaxivity determination; the principle requirements are that magnetic and radiofrequency fields, and gradient field linearity, be constant across the sample. A nuclear magnetic resonance (NMR) spectrometer could be used in place of an MRI scanner, though measurements would then generally need to be performed with one sample at a time.

4. Mutations are randomly introduced by the addition of $MnCl_2$ to the PCR. The mutation rate (average number of nucleotide substitutions per gene) can be tuned by altering the concentration of $MnCl_2$ in the reaction mixture. For most applications 1–2 nucleotide mutations per gene is ideal, which is here achieved by using a final concentration of 25 μM $MnCl_2$. Different mutation rates can be screened by varying ($MnCl_2$) from 5 to 200 μM, and assessing the fraction of functional variants.

5. The volume of transformed *E. coli* to be plated should be determined by the transformation efficiency of the competent cells and the number of variants to be screened. There must be enough colonies for the high-throughput screens, yet they need to be sparse enough to be individually picked. If unsure of the transformation efficiency, multiple transformation volumes can be plated on different plates. We typically screened 400–800 variants per generation of directed evolution.

6. We typically only screen one to two plates at a time because performing titrations on multiple neurotransmitters can be quite laborious. This means that each round of directed evolution requires screening across multiple days.

7. 5–10 neurotransmitter concentrations should be chosen, so the final concentrations are approximately logarithmically spaced and centered around the K_d of the parent. Serial dilutions of the neurotransmitter are an effective method to achieve accurately spaced dilutions.

8. Reading the absorbance from 350 to 500 nm for each well at each neurotransmitter concentration can be slow, depending on the plate reader. We only read absorbance every 10 nm over the interval from 350 to 500 nm.

9. Ligand-induced spectral changes may result in shifting of the BM3h absorbance peak (λ_{max}) toward either shorter or longer wavelengths. Blue-shifting of λ_{max} is characteristic of fatty acid binding to BM3h, which is accompanied by a transition of the heme iron from hexacoordinate to pentacoordinate geometry.

Red-shifting of λ_{max} is observed with monoamine ligands, and is thought to indicate direct coordination of the heme iron by the ligand.

10. When analyzing the data, keep track of the increases in volume upon titration. This significantly affects the total neurotransmitter and protein concentrations, which are inputs into the curve fitting. We perform data analysis using the Matlab software package (Mathworks, Natick, MA), but many alternatives are available.

11. The variants to be selected are those that provide the greatest improvement in affinity and specificity toward the neurotransmitter of interest.

12. Cells can also be lysed with an ultrasonic probe or a French press.

13. For the bulk titrations, we make one concentrated neurotransmitter solution for each ligand concentration to be tested and add small amounts of this solution to the cuvette.

14. Given the size of our standard radiofrequency coil, we typically perform experiments in bisected 384-well plates (Fig. 4a); only wells near the center of the plate are used because of field inhomogeneity effects closer to the plate edges. In a different scanner and coil, these requirements could be different.

15. We have used a TE of 10 ms and TR values of 73, 116, 186, 298, 477, 763 ms, 1.221, 1.953, 3.125, and 5.000 s. An inversion recovery pulse sequence or an alternative T_1-weighting method could be applied in place of the spin echo sequence (9), provided that a suitable analytical approach is applied to extract relaxation rates from the data. Geometric parameters of the imaging should be chosen to achieve desirable signal-to-noise ratio for the given sample size. We image a single 2-mm slice positioned midway through the sample heights, with a 16×8 cm in-plane field of view and a data matrix of 512×128 points.

16. We perform both image reconstruction and relaxation rate determination using custom routines running in Matlab. Equivalent analysis procedures could be implemented with other numerical data processing programs, however, including the software supplied with some MRI scanners.

17. In generating MRI dopamine sensors (6), we found that incorporation of a thermostabilizing mutation (10) improved our ability to continue directed evolution after the first four rounds. By increasing a protein's tolerance to mutagenesis, stability enhancements can facilitate directed evolution in general (11).

Acknowledgments

We are grateful to Robert Langer and to collaborators in the Arnold and Jasanoff laboratories for participation in the initial research and discussions that helped establish this protocol. We thank Eric Brustad for preliminary data used in some of the figures. MGS acknowledges the Fannie and John Hertz Foundation and the Paul and Daisy Soros Fellowship for support. This work was funded by NIH grant numbers R01-DA28299 and DP2-OD2441 (New Innovator Award) to A.J., and NIH grant number R01-GM068664 and a grant from the Caltech Jacobs Institute for Molecular Medicine to F.H.A.

References

1. Jasanoff A (2007) MRI contrast agents for functional molecular imaging of brain activity. Curr Opin Neurobiol 17:593–600

2. Gilad AA et al (2007) Artificial reporter gene providing MRI contrast based on proton exchange. Nat Biotechnol 25:217–219

3. McMahon MT et al (2008) New "multicolor" polypeptide diamagnetic chemical exchange saturation transfer (DIACEST) contrast agents for MRI. Magn Reson Med 60:803–812

4. Shapiro MG, Szablowski JO, Langer R, Jasanoff A (2009) Protein nanoparticles engineered to sense kinase activity in MRI. J Am Chem Soc 131:2484–2486

5. Iordanova B, Robison CS, Ahrens ET (2010) Design and characterization of a chimeric ferritin with enhanced iron loading and transverse NMR relaxation rate. J Biol Inorg Chem 15:957–965

6. Shapiro MG et al (2010) Directed evolution of a magnetic resonance imaging contrast agent for noninvasive imaging of dopamine. Nat Biotechnol 28:264–270

7. Dougherty MJ, Arnold FH (2009) Directed evolution: new parts and optimized function. Curr Opin Biotechnol 20:486–491

8. Omura T, Sato R (1964) The carbon monoxide-binding pigment of liver microsomes. I. Evidence for its hemoprotein nature. J Biol Chem 239:2370–2378

9. Bernstein MA, King KF, Zhou XJ (2004) Handbook of MRI pulse sequences. Academic, New York

10. Fasan R, Chen MM, Crook NC, Arnold FH (2007) Engineered alkane-hydroxylating cytochrome P450(BM3) exhibiting nativelike catalytic properties. Angew Chem Int Ed Engl 46:8414–8418

11. Bloom JD, Labthavikul ST, Otey CR, Arnold FH (2006) Protein stability promotes evolvability. Proc Natl Acad Sci U S A 103:5869–5874

12. Lewis DFV (2001) Guide to cytochrome P450 structure and function. Taylor & Francis, New York

INDEX

Matthew R. Banghart (ed.), *Chemical Neurobiology: Methods and Protocols*, Methods in Molecular Biology, vol. 995,
DOI 10.1007/978-1-62703-345-9, © Springer Science+Business Media New York 2013

Printed by Publishers' Graphics LLC
LSI20130405.15.01.48